T0214742

SpringerBriefs in Education

We are delighted to announce SpringerBriefs in Education, an innovative product type that combines elements of both journals and books. Briefs present concise summaries of cutting-edge research and practical applications in education. Featuring compact volumes of 50 to 125 pages, the SpringerBriefs in Education allow authors to present their ideas and readers to absorb them with a minimal time investment. Briefs are published as part of Springer's eBook Collection. In addition, Briefs are available for individual print and electronic purchase.

SpringerBriefs in Education cover a broad range of educational fields such as: Science Education, Higher Education, Educational Psychology, Assessment & Evaluation, Language Education, Mathematics Education, Educational Technology, Medical Education and Educational Policy.

SpringerBriefs typically offer an outlet for:

- An introduction to a (sub)field in education summarizing and giving an overview of theories, issues, core concepts and/or key literature in a particular field
- A timely report of state-of-the art analytical techniques and instruments in the field of educational research
- A presentation of core educational concepts
- An overview of a testing and evaluation method
- A snapshot of a hot or emerging topic or policy change
- An in-depth case study
- A literature review
- A report/review study of a survey
- An elaborated thesis

Both solicited and unsolicited manuscripts are considered for publication in the SpringerBriefs in Education series. Potential authors are warmly invited to complete and submit the Briefs Author Proposal form. All projects will be submitted to editorial review by editorial advisors.

SpringerBriefs are characterized by expedited production schedules with the aim for publication 8 to 12 weeks after acceptance and fast, global electronic dissemination through our online platform SpringerLink. The standard concise author contracts guarantee that:

- an individual ISBN is assigned to each manuscript
- each manuscript is copyrighted in the name of the author
- the author retains the right to post the pre-publication version on his/her website or that of his/her institution

More information about this series at http://www.springer.com/series/8914

Azra Moeed · Brendan Cooney

Language Literacy and Science

Enhancing Engagement and Achievement in Science

Azra Moeed (iD)
Victoria University of Wellington
Wellington, New Zealand

Brendan Cooney
Porirua College
Porirua, New Zealand

ISSN 2211-1921   ISSN 2211-193X (electronic)
SpringerBriefs in Education
ISBN 978-981-16-4000-1   ISBN 978-981-16-4001-8 (eBook)
https://doi.org/10.1007/978-981-16-4001-8

This Springer imprint is published by the registered company Springer Nature Singapore Pte Ltd.
The registered company address is: 152 Beach Road, #21-01/04 Gateway East, Singapore 189721, Singapore

Acknowledgements

We would like to thank the participating school, teachers, and students who have so willingly given their time and New Zealand Ministry of Education for providing the funding for this research. Our sincere thanks to Susan Kaiser and Dr. Abdul Moeed for reading the drafts and editing, and Irene Sattar for formatting the book for us.

About This Book

We set out to investigate reasons for low engagement and achievement in science, with the goal of finding ways to identify the cause and possible solutions through a research project. Making connections was a two-year project where all teachers in one secondary school used teaching as inquiry to research their own practice to enhance student's science learning. Teachers intentionally made use of literacy-based instructional strategies for teaching science, the primary objective being to support students using science words in talking, discussing, and writing. The research project was situated in a coeducational state secondary school in a low socio-economic area of New Zealand that has a very high percentage of students from Pasifika (Pacific islands) backgrounds (66%), and nearly half as many Māori (Indigenous New Zealanders (27%) and the rest were from other ethnicities. A salient feature of the project was that all participating teachers researched their own practice using a teaching as inquiry model. In this book we report the findings of two case studies of students' science learning. Data collection involved both qualitative and quantitative methods: student questionnaires; analysis of student work; students' assessment; and focus group interviews with students and with their teachers. This book presents findings that suggest that the language-enhanced learning strategies had a positive impact on students' science learning and achievement. As language literacy is a global issue in science education, the research offers accessible solutions to this problem of science education practice.

Keywords Science vocabulary development · learning · learning strategies · concept maps · teaching as inquiry · students view about science learning · EAL in science · indigenous science learners · enhancing achievement of Pasifika students in science; enhancing achievement of Māori students in science.

Contents

List of Figures

List of Tables

Chapter 1
Introduction and Literature Review

Abstract Internationally and in New Zealand, goals of twenty-first century school science include students developing conceptual understanding, procedural understanding, and an understanding about the Nature of Science, alongside the aim of developing scientifically literate citizens capable of making informed decisions about socio-scientific issues they encounter in their everyday lives (Bull, 2010; Hodson, 2014; Moeed & Easterbrook, 2016). The notion that developing an understanding of science requires the student to talk, read and write, that is literacy is the key focus of this research. "Developing an understanding of an idea requires talking about it, writing about it, reading about it and representing/drawing or visualizing it" along with engaging in practical work (Osborne, 2015, p. 18). The notion that developing an understanding of science requires the student to talk, read and write, that is literacy is the key focus of this research.

Keywords Language literacy and science literacy · Enhancing achievement of Pasifika students and Maori students in science · Science learning for EAL students

Introduction

Learning science is complex and successive Programme for International Student Assessment (PISA) results have shown that in general New Zealand students have performed well in science. However, Māori (indigenous New Zealanders) and Pasifika (those from Pacific island backgrounds) students have not performed so well in PISA. This trend continues with the most recent PISA assessment showing that although Pākeha (European New Zealanders) and Asian students have scored above the Organisation for Economic Cooperation and Development (OECD) average in Science, Māori and Pasifika students have scored below the OECD average (Ministry of Education, 2016). Fewer Māori and Pasifika students are studying Science and Mathematics in senior secondary school in New Zealand and their achievement is lower in Levels 1, 2 and 3 of the National Certificate of Educational Achievement (NCEA) and the few who continue with sciences beyond school show low levels of retention and achievement (Philp, 2014).

In the current global environment, which is reliant on information, innovation, and invention it is not only being successful in science but also in Science Technology Engineering and Mathematics (STEM) that matters. Current research shows that Pasifika and Māori students are underrepresented in all STEM subjects (Leggon & Gaines, 2017).

Language Literacy and Scientific Literacy

Language Literacy

Literacy skills are critical to building knowledge in science. Hicks, Hyler and Pangle (2020) draw attention to a lack of writing in science classrooms. This means students generally lack skills in basic science writing. Often this lack of writing skills may not be exposed until senior secondary levels when they have to write in examinations. Written work or book work in junior science class can ideally be used to teach the process of deciding what to write, how to write and what Hicks et al. (2020) call "to experience the painstaking work of moving from thought to words, and from words to sentences (p. 27). Several opportunities to practice lead to the student developing writing skills and they are able to see the resultant progress (Hicks et al. 2020). Having advanced skills in writing may well help students to write science assessments and achieve.

Hayden and Eades-Baird (2020) highlight that many science teachers have not been prepared with the skills to notice how literacy is integral to their science discipline. In science, students are expected to make observations, explain, obtain and use information from texts/media, argue, describe, ask questions, and communicate (Bull, 2015). Therefore, language and literacy are important to the 'doing' and communicating of science. This communicating ideas through oral and written language: reading, writing, and using discipline-specific texts; explaining; arguing; discussing; asking; and describing is only possible if the students have been taught how to do this and have opportunities to practice it (Wright & Domke, 2019).

Science Literacy

It is through communicating their science ideas that students provide evidence of their learning, whether it is through writing, demonstration, drawing, or verbal communication. When ways of communicating are limited due to poor literacy, it is a major issue and impacts on students' ability to express their developing knowledge and skills in science. There is a growing recognition that for students who have literacy issues, success in science is a real challenge and internationally there is a growing concern about this issue. As said earlier, this may be reflected in the achievement of these students when they are assessed during internal or external examinations.

There is a heightened awareness of the need for students to comprehend and communicate their growing science knowledge. Science education scholars Wellington and Osborne (2001) assert that:

- Learning the language of science is a major part (if not *the* major part) of science education. Every lesson is a language lesson.
- Language is a major barrier (if not *the* major barrier) to most pupils in science learning (p. 2).

Similarly, Watts (2003) has highlighted the need for students to be able to read, write, and speak their science learning. Markic and Childs (2016) highlight the demotivating effects of a lack of linguistic skills which constrain student's ability to ask questions, plan and conduct investigations and communicate it using the language of science. Further, that it is essential that students can comprehend and explain science ideas clearly, thus demonstrating their understandings of those ideas. Earlier, Lemke (1998) talked about the language of science being an integration of words, pictures, diagrams, charts, tables, and equations. *The New Zealand Curriculum* (Ministry of Education, 2007) requires development of the key competency of using symbols and texts along with thinking, managing self, relating to others, and participating and contributing.

More recently, in a commentary on literacy research in science education, Yore (2018) explores the historical change in focus in science education. For example, for a long period of time during the 1960s, science texts were encyclopaedias of knowledge which were inaccessible due to their high reading and comprehension requirements (Yore & Tippett, 2014). In the next decade, the focus in science education shifted to more *doing science* where inquiry learning became a common pedagogical approach (Waring, 1979). Inquiry-based learning is an active learning approach that starts by posing questions, problems, or scenarios. This contrasts with the traditional education, which generally relies on the teacher presenting facts and their own knowledge about the subject. Inquiry-based learning is often assisted by a facilitator rather than a teacher. This century, leading scholars for example, Abrahams (2009); Abrahams & Millar, (2008) assert that science learning is not just *hands-on* but requires a *minds-on approach* so that the students not only manipulate objects but also are able to manipulate associated science ideas. They highlighted the need for students to think about science, not just *do* science. However, the problem remains when a proportion of the population may well be able to do science and think about it but *cannot* write or talk about it due to language barriers. Nicholas and Fletcher (2017) argue that in an increasingly diverse population, as New Zealand now has, not being able to express their ideas in English is impacting on student achievement, not just in science but in other areas as well.

Blachowicz and Fisher (2007) states that although the importance of vocabulary and its effect on school achievement has been researched and reported but this has not informed teacher practice. She points out that there is a gap in vocabulary knowledge between students from higher and lower economic communities, and if this gap exists in early childhood it persists through the rest of their schooling years. One reason for this may be that children from disadvantaged backgrounds may have had less exposure to new words in their daily lives (Blachowicz & Fisher, 2007). McKinley (2005) explains that many "teachers in schools have been through a teacher education system that did not consider cultural differences in classrooms as part of pedagogy or saw the issue of indigenous student 'underachievement' as a deficit in the culture or the person" (pp. 230–31). González, Moll, and Amanti (2006) argue that teachers who have a deficit view tend to ignore the funds of knowledge that students with diverse cultures bring to the classroom. The idea of funds of knowledge is based on the premise that people are competent and have knowledge and this knowledge comes from their life experiences (González et al., 2006). The role of discourse was raised by Lee (1997) who argued that "students bring their own way of looking at the world that is representative of their cultural and language environments" (p. 221). Taber (2015) argues that some simple concepts that are experiential in nature could possibly be communicated by showing and labelling, but sophisticated communication is needed to express abstract ideas, and science has no shortage of such abstract ideas.

Osborne (2015) argues that Science is about a set of ideas and that the role of the science teacher is to help build students understanding of science ideas and how to understand the scientific practices that have led to the development of these ideas. Osborne suggests that developing understandings of these ideas "requires talking about it, writing about it, reading about it and representing/drawing or visualising it" (p. 18). Having the tools to talk write and represent science was of interest in this research project.

Added to the complexity of learning science is science education's preferred ways of assessing through pen and paper tests which are still the mode of assessment for international tests such as PISA. This was an issue for the participants of this research who, along with coming from a different cultural background, were linguistically diverse and lived in a low socio-economic community.

Scientific literacy is the "capacity to use scientific knowledge, to identify questions and draw evidence- based conclusions in order to understand and make decisions about the natural world and the changes made to it through human activity" (Harlen, 2001 p. 84). In the context of school science, Bybee and McCrae (2011) assert that scientific literacy is the student's ability to apply their knowledge and skills in novel settings, "because that is what they will have to do as citizens" (p. 8).

Bybee (1995) proposes the idea of dimensions of scientific literacy. These include: *1. Functional scientific literacy* (vocabulary, technical words of science and technology); *2. Conceptual and procedural scientific literacy* (relating information and experiences to the conceptual ideas that unify the various disciplines within science whereas procedural literacy is about the processes and procedures which is unique to scientific ways of knowing); *3. Multidimensional scientific literacy* (related to nature of science and technology and its role in personal life and society.

This notion of dimensions of scientific literacy provided a useful framework for the research reported here.

Not all students learn in the same way, so using different approaches to teaching and enabling students through developing learning strategies helps them to learn

(Moeed & Easterbrook, 2016). Most students in the research school were of Pasifika ethnicities and although they could speak English their language skills were limited. The second largest group in the sample was of Māori students some of whom were likely to be bilingual.

Science Learning Challenges for English as an Additional Language Learners

Language is part of who our students are and how they make sense of the world around them. Although teachers are told that language for learning is a central element in all teaching, student teachers get a cursory look at the issue in the over-crowded science teacher education programmes. In our own experience, we can say that student teachers of science during their training are not adequately prepared for the expectation of teaching in linguistically diverse, multi-lingual classrooms.

English as an Additional Language (EAL) learners bring their own funds of knowledge to the largely monolingual science class where all communication takes place in English. Although science teachers may not be prepared to provide targeted support to EAL learners they are expected to provide the linguistic support to help their EAL learners to succeed in science alongside their classmates who have English as their first language (Daborn, Zacharias & Crichton, 2020; Frantz, 2020).

Children with English as an additional language can benefit from targeted interventions. Although there is considerable research into EAL, there is a paucity of scientific evidence of effective intervention to support these children in their science learning. Oxley and De Cat (2019) reviewed language and literacy interventions with EAL students in the United Kingdom. They found that all the interventions in their review targeted vocabulary, either directly or indirectly (e.g. as part of a comprehension measure). They also claim that their results provide strong evidence that explicit vocabulary teaching and through targeting oral language practice has a positive impact for EAL learners. Thus far, few have investigated targeted specific aspects of grammar, for example, sentence structure all being in passive voice which pose difficulties for EAL children (Armon-Lotem, de Jong & Meir, 2015).

Science teachers in this research had multiple demands on them, and these would be similar in other countries with diverse ethnicities. Teachers have to prepare students, so they understand science ideas and be able to communicate them. The curriculum requires conceptual, procedural, nature of science understandings, and by the end of schooling for students to be scientifically literate. Parents want their children to be successful in school and beyond, and the school wants good achievement in national examinations. It becomes a matter of looking at the complex issue and deciding what to prioritize and what will in the short term engage students in learning activities and in the long run support students and teachers to meet the multifarious demands. During our research added to this was the challenge of student motivation to come to class and want to learn. Therefore, the focus of the research project was for students to use the scientific terminology (language of science) and to be able to demonstrate their understandings of the science ideas associated with scientific terminology through communicating it in talk, written format and through visual representation such as concept maps. The participating teachers had to consider the options and make some decisions about what to prioritise.

Teaching as Inquiry

Being successful in learning is motivational and when students experience success it has positive outcomes for their future learning (Midgely, 2014). In this research project the priority for the teachers was for students to experience success and become motivated to want to learn science, the biggest barrier to this being their inadequate science literacy and English literacy. *The New Zealand Curriculum* (Ministry of Education, 2007) promotes Teaching as inquiry, that has a focus on what the teacher does to be a better teacher and more effective in their teaching. There are three questions that guide this process depending on the what the teacher wants to improve (See Fig. 1.1). The curriculum explains why an inquiry approach to teaching is important: "Since any teaching strategy works differently in different contexts for different students, effective pedagogy requires that teachers inquire into the impact of their teaching on their students" (p. 35). Participating teachers were seeking to look at their teaching to have a positive impact on students' learning. Previously,

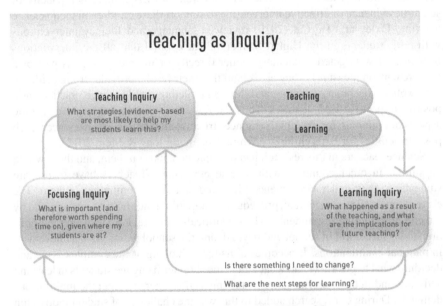

Fig. 1.1 Teaching as inquiry model (Ministry of Education, 2007, p. 35)

they had often tried out new strategies that they had read about or learnt through involvement in professional development. They had made intuitive decisions about whether the strategies had worked or not and these decisions were neither evidence based nor routinely shared with other members of their department or with teachers in other schools. Participating teachers wanted to change this practice and made a conscious effort to use Teaching as Inquiry model (Fig. 1.1). Individually and as a team, they wanted to establish a baseline, and prioritise enhancing student achievement in science classes in Years 9 (age 13 years). They decided to teach specific learning strategies to help students develop science specific vocabulary. Thereby supporting their students to become better learners and be able to communicate their science understandings.

Although some participating teachers had attempted to undertake Teaching as Inquiry, they appeared to have different understandings of what was required. They started from Focusing Inquiry. Time, guidance, and support were identified as requirements to research their practice and to achieve better outcomes for students. Participating in the research project provided the opportunity and the time for guidance and support. The teachers wanted students to be able to write paragraphs but realised that students did not have the words to do so. Subsequently, this led to the decision to focus on literacy learning strategies that would support students to develop vocabulary in the context of the topics they were studying and to make connections between science ideas.

The teachers decided to inquire into their practice and research was conducted alongside this to gain insight into how this teacher inquiry was conducted, the effectiveness of teacher inquiry as well as understanding what affords and constrains teachers in implementing it. There were clearly two aspects to this research, a practice focus and a research focus.

Practice Focus

Participating teachers in this project decided to inquire into their teaching with the view to improving their students' scientific vocabulary through using literacy enhancing strategies in their Year 9 (age about 13 years) science classes and by teaching and monitoring concept mapping to their Year 12 (age about 16 years) students. The research questions for teacher inquiry were:

- What learning strategies help year 9 students to develop science vocabulary in the context of their science learning?
- In what ways can teaching concept mapping can support Year 12 students to link science ideas and use these maps to write paragraphs to communicate their understanding of science ideas?

Research Focus

The research interest was to find out how teacher inquiry into practice was conducted, what was learnt, and to communicate it to those who may have similar issues. There were three questions for this research focus.

- How does targeted teaching and practice of literacy learning strategies influence Year 9 students' learning in science? (Findings are reported in Chap. 3)
- In what ways can concept mapping support Year 12 students to link science ideas and communicate these in writing? (Findings are reported in Chap. 4)
- What support and monitoring enable participating science teachers to inquire into their teaching with a view on enhancing student learning? (see Chap. 5)

In Chap. 2, we discuss the overall research design and methodology as well as the theoretical framework under which this research was conducted. Chapter 3 reports the *Year 9 Case study* and explains what *Sparklers* are and as they were used for developing science literacy. Sparklers are a collection of worksheets that are topic specific, designed at the accessible literacy level and put together as a booklet (See Appendix 1 for a sample). We explain the inquiry process, present evidence, and findings in relation to the first research question. Chapter 4 reports the *Year 12 Case study* and explains the concept mapping as in the previous chapter, we explain the inquiry process, present evidence, and findings in relation to the second research question. Chapter 5 reports the findings of *Teaching as inquiry* and will include the inquiry process, present evidence, and findings in relation to the third research question. Chapter 6 discusses the findings in the light of the extant literature evaluating the effectiveness of the learning strategies used in this research and the teacher inquiry. The chapter concludes with a commentry on the curriculum requirement of teachers researching their own practice and its affordances, constraints, and implications for practitioners.

References

Armon, S., de Jong, J., de Meir (Eds.). (2015). *Assessing Multilingual Children: Disentangling Bilingualism From Language Impairment*. Multilingual Matters.

Blachowicz, C. L., & Fisher, P. J. (2007). Best practices in vocabulary instruction. *Best Practices in Literacy Instruction, 3*, 178–203.

Bull, A. (2010). Primary science education for the 21st century: How, what, why? Retrieved from New Zealand Council for Educational Research, http://www.nzcer.org.nz/system/files/primary-scienceeducation-21st.pdf

Bull, A. (2015). *Capabilities for living and lifelong learning*. Wellington: NZCER.

Daborn, E., Zacharias, S., & Crichton, H. (2020). *Subject Literacy in Culturally Diverse Secondary Schools: Supporting EAL Learners*. Bloomsbury Publishing.

Frantz, K. K. (2020). Subject literacy in culturally diverse secondary schools: Supporting EAL learners. *Classroom Discourse*. https://doi.org/10.1080/19463014.2020.1797844

González, N., Moll, L. C., & Amanti, C. (Eds.). (2006). *Funds of Knowledge: Theorizing Practices in Households, Communities, and Classrooms*. Routledge.

Hayden, H. E., & Eades-Baird, M. (2020). Disciplinary Literacy and the 4Es: Rigorous and Substantive Responses to Interdisciplinary Standards. *Literacy Research: Theory, Method, and Practice.* https://doi.org/10.1177/2381336920937258.

Hicks, T., Hyler, J., & Pangle, W. (2020). *Ask, Explore, Write!: An Inquiry-driven Approach to Science and Literacy Learning.* Routledge.

Hodson, D. (2014). Learning science, learning about science, doing science: Different goals demand different learning methods. *International Journal of Science Education, 36*(15), 2534–2553.

Leggon, C., & Gaines, M. (2017). *STEM and Social Justice: Teaching and Learning in Diverse Settings.* Cham, Switzerland: Springer International. https://doi.org/10.1007/978-3-319-56297-1

Lemke, J. (1998). Multiplying meaning. *Reading Science: Critical and Functional Perspectives on Discourses of Science,* 87–113.

Markic, S., & Childs, P. E. (2016). Language and the teaching and learning of chemistry. *Chemistry Education Research and Practice, 17*(3), 434–438.

Ministry of Education. (2007). *The New Zealand Curriculum.* Wellington: Learning Media.

Ministry of Education. (2016). *PISA 2015: New Zealand headline results.* Retrieved from: http://www.educationcounts.govt.nz/publications/series/PISA/pisa-2015/pisa-2015-new-zealand-headline-results

Moeed, A., & Easterbrook, M. (2016). Promising teacher practices: Students views about their science learning. *European Journal of Science and Mathematics Education, 4*(1), 17–24.

Osborne, J. (2015). Practical work in science: Misunderstood and badly used. *School Science Review, 96*(357), 16–24.

Oxley, E., & De Cat, C. (2019). A systematic review of language and literacy interventions in children and adolescents with English as an additional language (EAL). *The Language Learning Journal,* 1–23.

Philp, M. (2014). Developing diversity. *Engineering Insight, 15*(4), 38–40.

Taber, K. S. (2015). Meeting educational objectives in the affective and cognitive domains: Personal and social constructivist perspectives on enjoyment, motivation and learning chemistry. *Affective dimensions in chemistry education* (pp. 3–27). Springer.

Waring, M. (1979). Background to Nuffield science. *History of Education, 8*(3), 223–237.

Watts, M. (2003). Language and literacy in science education. *Australian Science Teachers Journal, 49*(2), 43.

Wellington, J., & Osborne, J. (2001). *Language and Literacy in Science Education.* McGraw-Hill Education (UK).

Wright, T. S., & Domke, L. M. (2019). The role of language and literacy in K-5 science and social studies standards. *Journal of Literacy Research, 51*(1), 5–29.

Yore, L. (2018) Commentary on the Expanding Development of Literacy Research in Science Education. In: Tang KS., Danielsson K. (eds) *Global Developments in Literacy Research for Science Education.* Springer, Cham. https://doi.org/10.1007/978-3-319-69197-8_22

Yore, L. D., & Tippett, C. D. (2014). Reading and science learning. *Encyclopaedia of Science Education,* 821–828.

Chapter 2
The Theoretical Frame, Research Design and Methodology

Abstract In this chapter we present the theoretical frame, the research design, and over all methodology. The research was underpinned by the social constructivist theory of learning. The research paradigm used was action research. Action research was appropriate as it uniquely allows research and practice to not only coexist but to simultaneously work together in problem solving, as is the case in the research presented here.

Keywords Social constructivism · Case study research methodology

Theoretical Framework

The research was underpinned by the constructivist theory of learning. Constructivism provides a framework for thinking about the ways in which learners engage and make sense of the objects around them (Ferguson, 2007). This view of learning within science education has underpinned science educational research since the 1980s and continues to the present time (Driver & Bell, 1986; Driver et al., 1994; Duit, 2016; Taber, 2017). There have been critiques of constructivism as a philosophy (epistemology) that are beyond the scope of this book (see for example, Matthews, 1992, 1993, 2002, 2014). Constructivism maintains that the knowledge is constructed by the learner based on their existing ideas. In a constructivist pedagogical approach, the teacher gives primacy to finding out what the learner already knows and engaging them in learning activities that support them to make sense of what they already know. There are a variety of constructivist approaches that have been theorised including: radical or personal constructivism (Von Glasersfeld, 1995, 2012); social constructivism (Amineh & Asl, 2015; Solomon, 1995); contextual constructivism (Cobern, 1993, 2012a, b) amongst others. In this book we are concerned with constructivism in relation to teaching and learning, and in particular, social constructivism.

Solomon (1987) suggested that the social setting "makes an essential difference to the learning situation, to how the task is perceived and even to the tools for thought that will be used" (p. 63). There are two main elements of social constructivism; firstly, that human beings rationalise their experiences in working out how the world works, and secondly, that language has an essential role in the construction of reality (Leeds-Hurwitz, 2012). Learning is affected by students' reflection on their experiences, as well as from the reaction of others when they share their ideas (Moeed & Easterbrook, 2014).

From a pedagogical perspective, Baviskar, Hartle, and Whitney (2008) state that there are four features of a pedagogy based on constructivism. These include "eliciting prior knowledge, creating cognitive dissonance, application of new knowledge with feedback and reflecting on learning" (p. 4). When eliciting prior knowledge, Baviskar et al. (2008) remind teachers that the activity used to find out what the learner knows needs to be related to new knowledge to be learnt. Although the task used for new learning needs to be problematic and should require thinking about by the learner in this case, simple and achievable tasks were a better starting point to ensure engagement. Opportunities for showing the learner that learning has occurred can be through quizzes, group discussion, and presentations where students share their ideas with their peers. Finally, reflection can be in the form of formal or informal time at the end of a lesson for student to think and write down what has been learnt. Windschitl (2002), advises that it is important to make the learner aware that learning has taken place.

The participants were the four teachers, John, Tim, Bev, and Sue (pseudonyms) and their classes over two years. It is important to mention here that teacher inquiry into their practice was not linked to teacher performance appraisal in this project.

Research Design and Methodology

The research paradigm used was action research. Action research was appropriate as it uniquely allows research and practice to not only coexist but to simultaneously work together in problem solving. In New Zealand, a commonly mentioned reason for young people wanting to be teachers is to make a difference to the lives of those they teach (Moeed & Easterbrook, 2014). Supporting students to learn so that they will experience educational success is therefore at the heart of teaching practice. The participants of this research project identified a problem-of-practice and collectively decided to make changes to their practice. The investigation into teacher inquiry into their teaching took an action research approach. Action research is not a new paradigm, it was first established by Kurt Lewin in the 1940s (Bradbury, Lewis, & Embury, 2019). Corey (1953) argues that "action research is undertaken by practitioners in order that they may improve their practices' (p. 141). Action research places a powerful tool in the hands of the teacher and is transformative social learning with an agenda for change (Judah & Richardson, 2006). Calvert and Sheen (2015) explain that this approach to research involves four stages; "(1) identify a problem or question; (2) carry out an action; (3) observe and reflect on the outcome; and (4) plan another action" (p.227). The teaching as inquiry model followed by the teachers in this research (see Fig. 1.1) has similar stages although it is a cyclic model suggesting continuity.

Educational action research is often seen as emancipatory, it is practical and leads to knowledge building. However, if action research is to contribute to a knowledge base it must be open to critique by the research community, and to withstand critique it ought to be rigorous (Newton and Burgess, (2016). Critics of action research for example, Brown and Jones (2003) raise questions about what counts as improvement. "How can the researcher both "observe" reality as well as being part of it and thus be implicated in its continual creation and recreation"? (p.5). The teachers and the researcher in this research were cognisant of the dual roles.

If the purpose is to transform science education through innovation, research, reflection and revision, then action research is a promising strategy for science teacher education which is embedded in research. Mamlok-Naaman, Eilks, Bodner and Hofstein (2018) suggest that action research provides an evidence-based approach to improving science education practice and affords a different perspective on the research to practice relationship as was the case in this research project.

As action research is often reported as narrative inquiry, Heikkinen et al. (2012) challenge us to consider:

What is the relationship between interventions in the social and physical reality, which are sometimes called 'action research', and narrative accounts about those interventions? In other words, what is the relationship between social practice and language? (p. 5).

Heikkinen et al. (2012) discuss the five principles: historical; continuity; reflexivity; dialectics; workability; and evocativeness with respect to the validity issues raised in relation to action research. A detailed account of these is beyond the scope of this book and readers interested in deeper understandings of the issues may find Heikkinen et al.'s article both thought provoking and useful. In the context of the research presented here the notions of credibility and trustworthiness were considered essential and were considered at all stages of data collection, analysis, interpretation, and reporting.

This research took a case-study approach. Jones and Baker (2005) argue that there are few classroom-based case studies that provide detailed and in-depth understanding of the multiple variables of the classroom environment and student–teacher interactions and learning and suggest such case studies are needed to enhance our understanding of science learning in the classroom. Case studies provide "thick, rich descriptions of the phenomenon under study" (Stake, 1995, p. 42; Stake 2005) and use an inductive mode of reasoning from which "generalizations, concepts, or hypotheses emerge from an examination of the data grounded in the context itself" (Merriam, 1998, p. 13). As cases have boundaries, in this research the case was science teachers in one school inquiring into their practice with the view of enhancing student learning.

Multiple data sources were used to ensure evidence-based data and to be able to triangulate the data. These included Science Thinking with Evidence (STwE) Test at beginning of Year 9 for all students and at the end, diagnostic assessment at the start of the topic. review of Sparklers, student and teacher cogenerative (Cogens), students survey, and common tests. To include a reflective process cogenerative dialogues were planned (Roth & Tobin, 2002). Cogens a process of critical pedagogy and help to identify contradictions that might be changed with the goal of improving the quality of teaching and learning. This was to give the teachers and students the opportunity to talk about the classroom practices being used and how these might afford or constrain engagement with scientific ideas "…all participants in cogenerative dialogs are encouraged to speak their minds, identify specific examples to illustrate where improvements can be made, and also identify examples of exemplary practices or counter examples of those that exemplify a need to change" (Tobin, 2014, p. 181). However, we found that Cogens was not a culturally appropriate tool for data collection for the participating students as they did not freely communicate to teachers what was working for them and what was not in a small group situation. Pasifika students are respectful to their teachers and telling teachers what teaching approach was not working for them would be disrespectful. Consequently, after one attempt by most participating teachers the methodology was modified. A former student who was capable and well-liked by the student community was employed to do focus group interviews with the students. The audio recordings were transcribed and used for teacher reflection to decide on the changes that were needed in their teaching approach in the following cycle. The results of these interviews are reported in Chap. 3.

During full research team meetings, these were reflected upon and in the discussion that followed, changes to be made in the next inquiry cycle were agreed upon. For example, following the discussion that students were not comfortable being interviewed by the teachers, it was decided that a former student would be engaged to do this. The data were coded, and the codes are indicative of their responses.

Researcher Role

The researcher and the first author is an academic who researches science teaching and learning and was the research advisor on the project. She provided guidance to the teachers in designing the research project, helping in selecting data sources for the teacher inquiry, analysis of the data, quality control, for example, checking milestone reports written by the lead teacher. The researcher has a constructivist view of learning which has been presented as the theoretical frame for the research.

The analysis of interview data was done through a process of constant comparison (Merriam, 2001). Open coding which was completed jointly by the researcher and lead teacher to examine changes in teacher thinking and practices of supporting their students' learning. These codes were combined into more general categories (Miles & Huberman, 1994). If the appropriate coding for a response was not immediately apparent, it was compared with other responses by the participant and with those coded previously until agreement was reached by both researcher and the senior teacher. The unit of analysis was a phrase which conveyed meaning and could be allocated to more than one category, for example; are able to write a word correctly, can complete sparkler exercises by themselves, works with another student to complete the sparkler tasks, beginning to make appropriate use of science idea.

Research Questions and Sources of Evidence

We considered clarity in how the research questions would be answered was important. Table 2.1 shows the research questions and the sources of evidence used. For example, to answer the first research question related to the Year 9 case study, we began the data collection with a pre-test which is Science Thinking with Evidence (STwE) test produced, validated, marked and graded by New Zealand Council for Educational Research (NZCER). Students created a mind map for diagnostic assessment before each topic. All participating teachers reviewed the Sparklers in the middle of the topic. Students were surveyed to gain insight in their view about the usefulness of Sparklers for their learning. At the end of each topic, Sparklers were reviewed by the teachers. All participating students completed a topic test at the end of each topic. Finally, the STwE test was repeated at the end of the case study data collection.

To answer the second research question related to the Year 12 case study, the data collection began with students drawing concept maps to demonstrate their current knowledge of the topic. Students were surveyed to find out their views about the usefulness of concept maps for their learning. Students end of topic concept

Table 2.1 Research question and sources of evidence

Research questions	Sources of evidence
• How does targeted teaching and practice of learning strategies influence Year 9 students' achievement in science?	Year 9 Case Study 1: • Science Thinking with Evidence (STwE) Test at beginning of Year 9 for all students • Diagnostic assessment at the start of the topic • Review of Sparklers by teachers' mid cycle • Survey of students' views about the usefulness of Sparklers for their *learning* • Review of completed Sparklers at the end of the topic • Common topic tests for all year 9 classes • *Repeat with next topic* • STwE Test at end of Year 9 for all students
• In what ways can concept mapping support year 12 students to link science ideas and communicate these in writing?	**Year 12 Case Study 2:** • Students' initial concept maps • Survey students' views about the usefulness of this strategy for their "learning." • Students' final concept maps • Mock exam results for the relevant achievement standard
• What support and monitoring enables participating science teachers to inquire into their teaching with a view on enhancing student learning?	• Audio recordings of meetings with the team leader • Audio recordings of meetings with the group • Milestone reports • Teacher interviews

maps were collected. Students mock examination results for the relevant topics were collected.

As already noted, this involved four science teachers inquiring into their teaching practice with the view of helping students to develop their science vocabulary literacy skills, and ability to communicate their ideas in writing. The next three chapters present the results of this research.

Data Analysis

It was important that the interpretations represented the participants view and not those of the researcher (Fusch et al. 2018). For this reason, the data were analysed collaboratively to ensure that interpretations agreed with those of the teachers. In addition, to increase rigour multiple data points were used and findings presented to the teachers to check. The data from multiple sources was then triangulated (Denzin, 2009). To maintain consistency in interpretation constant comparison to the previous cycle for the same class was made.

References

Amineh, R. J., & Asl, H. D. (2015). Review of constructivism and social constructivism. *Journal of Social Sciences, Literature and Languages, 1*(1), 9–16.

Bradbury, H., Lewis, R.E., & Embury, D.C. (In press). Education Action Research: With and for the Next Generation. In: Mertler, C.A. (Eds.), *The Wiley Handbook of Action Research in Education.* Hoboken, NJ: John Wiley & Sons

Calvert, M., & Sheen, Y. (2015). Task-based language learning and teaching: An action-research study. *Language Teaching Research, 19*(2), 226–244.

Corey, S. M. (1953). *Action Research to Improve School Practices.* Bureau of Publications.

Cobern, W. W. (2012a). Contextual constructivism: The impact of culture on the learning and teaching of science. *The Practice of Constructivism in Science Education* (pp. 67–86). Routledge.

Cobern, W. W. (2012b). Contextual constructivism: The impact of culture on the learning and teaching of science, The National Association for Research in Science Teaching, Lake Geneva, Wisconsin, April 7–10, 1991.https://doi.org/10.3102/0013189X023007005

Denzin, N. K. (2009). *The Research Act: A Theoretical Introduction to Sociological Methods* (3rd ed.). Prentice Hall.

Driver, R., Asoko, H., Leach, J., Scott, P., & Mortimer, E. (1994). Constructing scientific knowledge in the classroom. *Educational Researcher, 23*(7), 5–12. https://doi.org/10.3102/0013189X023007005

Duit, R. (2016). The constructivist view in science education: What it has to offer and what should not be expected from it. *Investigações Em Ensino De Ciências, 1*(1), 40–75.

Eilks, I., Frerichs, N., & Kapanadze, M. (2018). Action Research to Innovate Science Teaching-An ERASMUS+ Capacity Building Action in Science Teacher Education19.

Ferguson, R. L. (2007). Constructivism and social constructivism. In G. M. Bodner & M. Orgill (Eds.), *Theoretical Frameworks for Research in Chemistry/Science Education* (pp. 28–49). Upper Saddle River, NJ: Pearson Prentice Hall.

Fusch, P., Fusch, G. E., & Ness, L. R. (2018). Denzin's paradigm shift: Revisiting triangulation in qualitative research. *Journal of Social Change, 10*(1), 2.

Heikkinen, H. L., Huttunen, R., Syrjälä, L., & Pesonen, J. (2012). Action research and narrative inquiry: Five principles for validation revisited. *Educational Action Research, 20*(1), 5–21.

Jones, A., & Baker, R. (2005). Curriculum, learning and effective pedagogy in science education for New Zealand: Introduction to special issue. *International Journal of Science Education, 27*(2), 131–143.

Judah, M. L., & Richardson, G. H. (2006). Between a rock and a (very) hard place: The ambiguous promise of action research in the context of state mandated teacher professional development. *Action Research, 4*(1), 65–80.

Leeds-Hurwitz, W. (2012). These fictions we call disciplines. *The Electronic Journal of Communication, 22*(3–4).

Mamlok-Naaman, R., Eilks, I., Bodner, G., & Hofstein, A. (2018). *Professional Development of Chemistry Teachers: Theory and practice*. Royal Society of Chemistry.

Matthews, M. R. (1992). Constructivism and empiricism: An incomplete divorce. *Research in Science Education, 22*(1), 299–307. https://doi.org/10.1007/BF02356909

Matthews, M. R. (1993). Constructivism and science education: Some epistemological problems. *Journal of Science Education and Technology, 2*(1), 359–370. https://doi.org/10.1007/BF0069 4598

Matthews, M. R. (2002). Constructivism and science education: A further appraisal. *Journal of Science Education and Technology, 11*(2), 121–134. https://doi.org/10.1007/978-94-011-5032-3

Matthews, M. R. (2014). *Science Teaching: The Contribution of History and Philosophy of Science*. Routledge.

Merriam, S. B. (2001). Case studies as qualitative research. *Qualitative Research in Higher Education: Expanding Perspectives, 2*, 191–201.

Merriam, S. B. (1998). *Qualitative Research and Case Study Applications in Education. Revised and Expanded from" Case Study Research in Education."*. Jossey-Bass Publishers, 350 Sansome St, San Francisco, CA 94104.

Miles, M. B., & Huberman, A. M. (1994). *Qualitative Data Analysis: An Expanded Sourcebook*. sage. https://doi.org/10.1080/0950069042000276686

Moeed, A., & Easterbrook, M. (2014). Science teacher development through partnership with a mentor: Eight years of teaching and learning. *International Journal of Professional Development, 3*(1), 3–16.

Newton, P., & Burgess, D. (2016). Exploring types of educational action research: Implications for research validity. In *The Best Available Evidence* (pp. 33–46). Brill Sense.

Roth, W. M., & Tobin, K. (2002). Redesigning an" urban" teacher education program: An activity theory perspective. *Mind, Culture, and Activity, 9*(2), 108–131.

Solomon, J. (1995). Constructivism and quality in science education. *Subject Learning in the Primary Curriculum: Issues in English, Science and Maths, 137*.

Solomon, J. (1987). Social influences on the construction of pupil's understanding of science. *Studies in Science Education, 14*, 63–82.

Stake, R. E. (1995). *The Art of Case Study Research*. SAGE Publications.

Stake, R. E. (2005). Qualitative case studies. In: Denzin, N. K., & Lincoln, Y. S. (Eds.), *The Sage Handbook of Qualitative Research* (p. 443–466). Sage Publications Ltd.

Windschitl, M. (2002). Framing constructivism in practice as the negotiation of dilemmas: An analysis of the conceptual, pedagogical, cultural, and political challenges facing teachers. *Review of Educational Research, 72*(2), 131–175.

Chapter 3
Year 9 Case Study, Sparklers
for Developing Science Literacy

Abstract This Chapter presents the Year 9 Case study. The focus of the case study was to help students to learn the language of science, understand the science ideas, and be able to communicate these in writing. The participating teachers collectively decided that they wanted to conduct this study with their Year 9 classes. This was because they wanted to have the classes who had learnt these strategies to have the opportunity to consolidate them in Year 10. This was made possible by the school allowing the participating teachers to have the same classes in the following year.

In Year 9 case study teachers used two teaching strategies.

- Sparklers and weekly quizzes.
- Supporting students to communicate their science ideas through writing

Keywords Strategy for developing science literacy · Science vocabulary development · Evidence of student learning

Sparklers are designed to reinforce what students are learning within a topic (see Appendix A). The activities include simple tasks to support vocabulary development, through to more challenging tasks that require analysis, synthesis, and application of what they are learning. Sparklers were developed by a teacher and were used with Year 9, 10, and 11 science classes in a high decile school where they are reported to have a positive impact on students' learning. A weekly written quiz was conducted in each class to check students' understandings of the ideas associated with the vocabulary learnt during the week, for sample, quiz questions (see Appendix B). These quizzes were carried out to ascertain what had been learnt and what needed further clarification. Figure 3.1 shows a diagrammatic representation of the process followed for the Year 9 case study.

Results for the Year 9 Teacher Inquiry

In this section we summarise the findings from the evidence collected from Year 9 interviews, surveys, pre and post-mind maps, and test results.

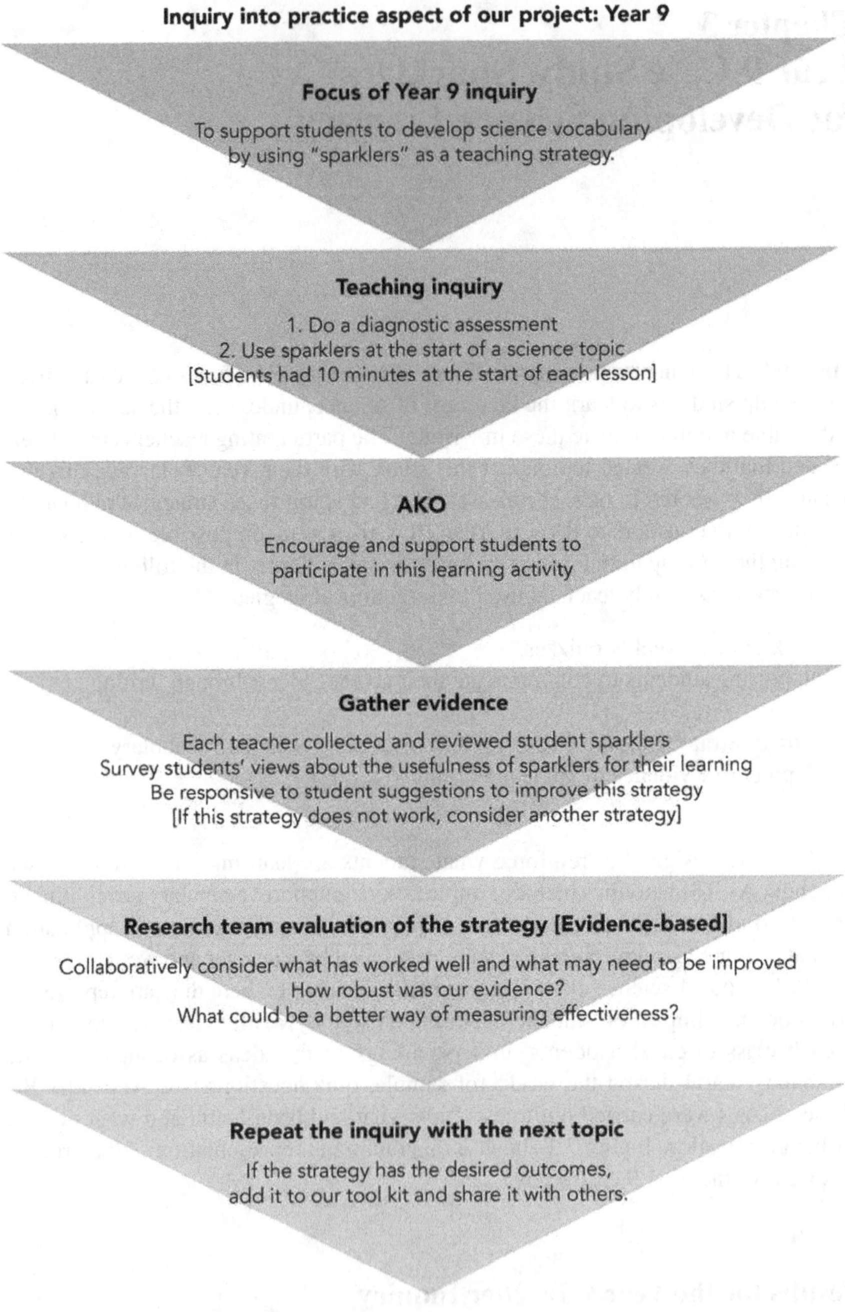

Inquiry into practice aspect of our project: Year 9

Focus of Year 9 inquiry

To support students to develop science vocabulary
by using "sparklers" as a teaching strategy.

Teaching inquiry

1. Do a diagnostic assessment
2. Use sparklers at the start of a science topic
[Students had 10 minutes at the start of each lesson]

AKO

Encourage and support students to
participate in this learning activity

Gather evidence

Each teacher collected and reviewed student sparklers
Survey students' views about the usefulness of sparklers for their learning
Be responsive to student suggestions to improve this strategy
[If this strategy does not work, consider another strategy]

Research team evaluation of the strategy [Evidence-based]

Collaboratively consider what has worked well and what may need to be improved
How robust was our evidence?
What could be a better way of measuring effectiveness?

Repeat the inquiry with the next topic

If the strategy has the desired outcomes,
add it to our tool kit and share it with others.

Fig. 3.1 The process of Year 9 Case study. *Note* AKO is a reciprocal learning relationship *Note* AKO is a reciprocal learning relationship. In te ao Māori, the concept of AKO means both to teach and to learn. It recognises the knowledge that both teachers and learners bring to learning interactions, and it acknowledges the way that new knowledge and understandings can grow out of shared learning experiences

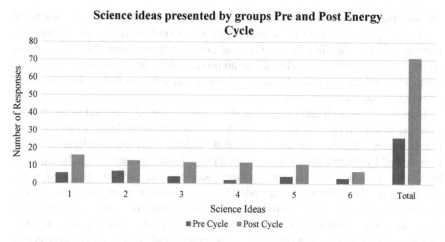

Fig. 3.2 Comparison of science ideas presented by students in assessment

Analysis of the Mind Maps Before and After the Energy Unit

The first cycle of the project started in Term 3 2015. In the first cycle, two classes were selected (taught by experienced teachers) so that any issues could be resolved before including the other teachers. The first topic was 'Energy'.

The first cycle showed that participation in the use of 'Sparklers' was high. Perhaps this was due to their novelty and students being able to experience success in completing the tasks. Some students needed more support than others, but it was a positive start. The before and after mind maps showed an improvement in Science ideas used by students (see Fig. 3.2).

Students constructed mind maps in groups at the start and end of the unit. The responses listed below are from counting each science idea presented on the group task sheet. We identified three misconceptions in the pre-test and one of these persisted in the post-test. The data presented in Fig. 3.2 are from six groups.

In the second cycle four classes were involved, in school term four 2015, the four participating teachers did a pre-topic mind map task and a post-topic map. The two classes involved in both cycles completed self-evaluation and teacher feedback forms, student interviews were conducted.

Results of Year 9 Student Interviews

We found that Cogens was not a culturally appropriate tool for data collection for the participating students as they did not freely communicate to teachers what was working for them and what was not in a small group situation. Pasifika students are respectful to their teachers and telling teachers what teaching approach was not working for them would be disrespectful. Consequently, after one attempt by most participating teachers the methodology was modified. A former student who was capable and well-liked by the student community was employed to do focus group interviews with the students. The audio recordings were transcribed and used for teacher reflection to decide on the changes that were needed in their teaching approach in the following cycle.

During full research team meetings, these were reflected upon and in the discussion that followed, changes to be made in the next inquiry cycle were agreed upon. For example, following the discussion that students were not comfortable being interviewed by the teachers, it was decided that a former student would be engaged to do this. The data were coded, and the codes are indicative of their responses. For example, the following were coded as group work: working in pairs, ask my neighbour, bouncing ideas around with friends, working in groups, group discussions, asking my friends for help, and working with friends. Similarly, working on a laptop, using Google, doing research on computers, and chrome books were coded as computers.

A few codes had single responses which included: being able to choose topic, scaffolding is helpful, and reading. One student said, "posters are a waste of time; tests show learning, so why a poster too?" Five students said that Sparklers were easy, and the majority found them useful, helpful, and fun to do. The results are presented in Fig. 3.3.

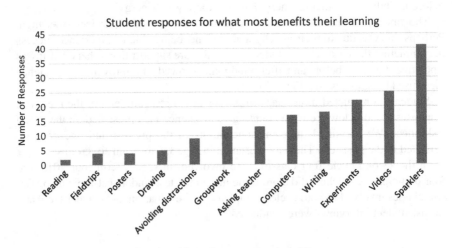

Fig. 3.3 Summary of Year 9 students' interview responses (n = 60)

Table 3.1 Results of the Year 9 survey

What helps them learn	Number of responses
Homework	30
Challenging tasks	40
Worksheets	42
Group competition	44
Copying down	47
Working alone	55
Group discussion	63
Listening to music	68
Doing experiments	72
Working in pairs	75
Using a computer	79

Results of Year 9 Student Surveys

Year 9 students completed a survey at the end of the study. The results are presented in Table 3.1. The responses show that students believed using the computers, doing experiments, working together, and listening to music most benefitted their learning. Activities such as worksheets (i.e., Sparklers) were identified as useful, but to a lesser extent.

Sparklers

In most classes, students enjoyed doing Sparklers for the first 10 min of the lesson. They worked cooperatively and helped each other and in the words of one teacher, "The Sparklers were addictive" and another said that they were helpful with students learning the words they needed to learn. A few students said that they found the Sparklers easy.

Year 9 students also completed a vocabulary quiz at the start and end of each topic which was based on the words they were learning through the Sparklers. A sample of the results from the first pre- and post-vocabulary tests of the year are presented in Fig. 3.4. During the year students became familiar with the learning strategies, and student vocabulary progressively improved.

There was a range in the completion of the Sparklers. Some students completed all the tasks and most managed most of the tasks. A few students struggled but did as much as they could on their own but sometimes sought support from their peers.

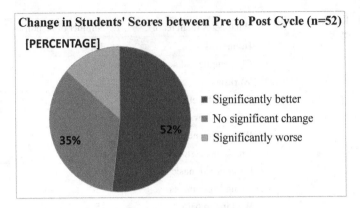

Fig. 3.4 Results of a sample pre- and post-vocabulary test for Space Cycle

Results from the pre- and Post-Mind Maps in the Second Cycle

At the start of each unit, students wrote the words they were familiar with that were related to the topic and this was repeated at the end of each topic. Figures 3.5, 3.6, and 3.7 show the percentage of mind maps containing the keywords at the start and end of the unit on Forensics and Ecology. The Forensic mind maps show a focus on the inquiry process that was emphasised in this unit, whereas the Ecology mind

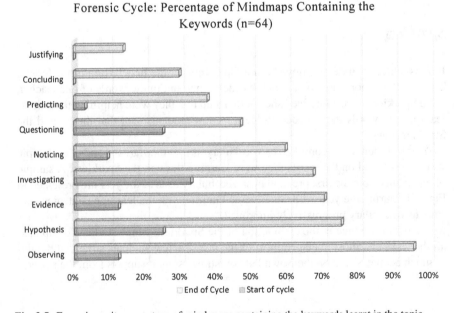

Fig. 3.5 Forensics unit, percentage of mind maps containing the keywords learnt in the topic

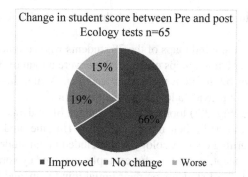

Change in student score between Pre and post Ecology tests n=65

15%

19%

66%

■ Improved   ■ No change   ■ Worse

Fig. 3.6 Change in student scores between pre and post Ecology test

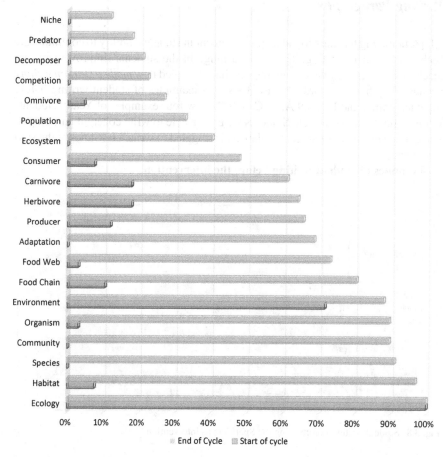

Ecology: Percentage of Mindmaps Containing the Keywords

■ End of Cycle   ■ Start of cycle

Fig. 3.7 Ecology unit, percentage of mind maps containing the keywords learnt in the topic

maps reflect the nature of the topic for which the students had to learn a vast number of new words.

When compared, the mind maps of the 78 students who completed both the pre- and post-cycle ecology mind maps exercise, there were a total of 60 different ideas before and a total of 69 different ideas after teaching. Again, this is a rather blunt measure, and does not provide a lot of useful information.

The graph above (Fig. 3.7) looks at the percentage of students who included the keyword (or a form of it) in their mind map before (in blue) and after (in orange) the teaching and learning cycle. Ecology was included on all student mind maps as it was the central concept. So, we see for example, that the keyword *food chain* was included by about 10% of students on their initial mind map, and by about 80% of the students on their final mind map.

Writing Paragraphs

All participating teachers found an improvement in students' ability to communicate both orally and in writing. There was a range in the students' improvement, but teachers said that even those students who had struggled now *made a start on writing* because they had the words. Figure 3.8 shows examples of student writing before the intervention and Fig. 3.9 A, B, C, and D show four examples of Year 9 students writing after the intervention. Students have been able to write coherent paragraphs, whereas according to their teacher they had struggled to write a few lines at the start of the unit.

Examples of student writing before the intervention.

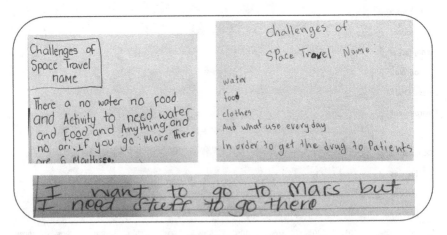

Fig. 3.8 Student writing before using the Sparklers as intervention

Examples of student writing after the Sparklers.

The analysis of student writing after the intervention shows both evidence of researching the issues and the complex sentence structures as illustrated in Fig. 3.9.

A: *First sentence is complex, use of punctuations. Contrasting intense heat and unheated*

blackness. Understanding of science idea and its consequence is communicated.

Reason for leaving Earth:
Natural Natural disasters, Terrovist   A

Below is space for you explain and discuss the ideas you stated above. Make sure you refer back to your assessment task sheet to meet all the requirements.

Whod r In order to Stay alive in Space,
people need lots of things: food, oxygen,
Shelter an and, perhaps most importantly fuel
Its gonna be hard to Store that a big amount
of fuel on the Space Ship. Object in low -
Earth orbit (the Place you'd Park your mars
Space Ship While you built it) travel around
the world exe every 90 minutes. Druving half
that time, they experience the intense heat
of the Sun and then the unheated blackness
of Space.

Unless tanks are regularly Vented, containers
holding these materials are liable to explode

B: *Written as a report. Claims that Kepler is about the same size as Earth. The idea is*

reinforced by claiming 385 days to orbit its star. Provides an imaginary distance of

1,400 lightyears.

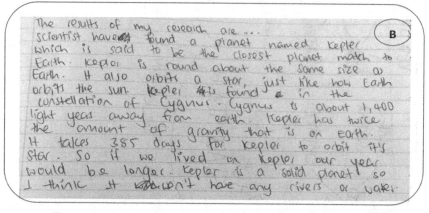

The results of my research are ...
Scientist have found a planet named kepler
which is said to be the closest planet match to
Earth. Kepler is round about the same size as
Earth. It also orbits a star, just like how Earth
orbits the sun. kepler is found a in the
constellation of Cygnus. Cygnus is about 1,400
light years away from earth. Kepler has twice
the amount of gravity that is on Earth.
It takes 385 days for kepler to orbit it's
Star. So if we lived on kepler our year
would be longer. Kepler is a solid planet so
I think it won't have any rivers or water.

Fig. 3.9 A, B, C and D are sample of a Year 9 student's writing after the intervention

C: *A complex first sentence, eluding to the risk. Then presents a justified reason for trying.*
        The last reason presents an opposing idea that we do not leave but make the Earth a
        better place.

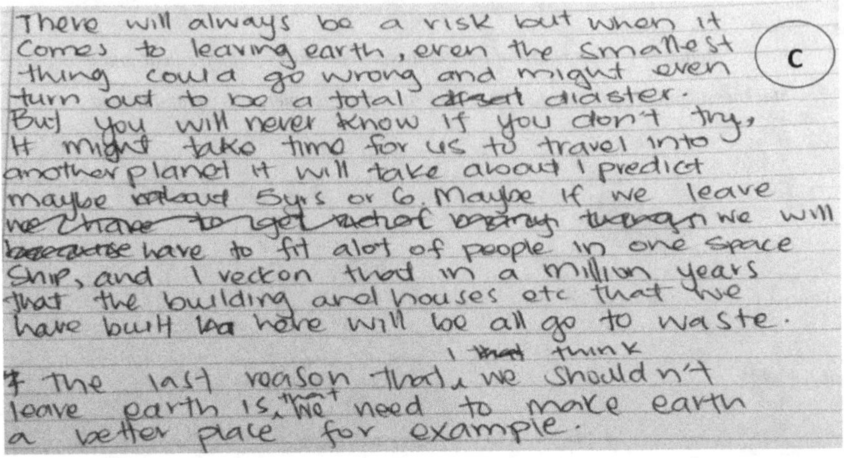

There will always be a risk but when it
comes to leaving earth, even the smallest
thing could go wrong and might even
turn out to be a total disaster.
But you will never know if you don't try,
It might take time for us to travel into
another planet it will take about I predict
maybe about 5yrs or 6. Maybe if we leave
we there to get another boring through we will
because have to fit alot of people in one space
ship, and I reckon that in a million years
that the building and houses etc that the
have built here will be all go to waste.
I think
The last reason that we shouldn't
leave earth is, "We" need to make earth
a better place for example.

C

D: *Using evidence to justify the possible danger. Citing reliable source for the depletion of*
        ozone. Explains the relationship between UV rays and cancer and how to protect
        oneself.

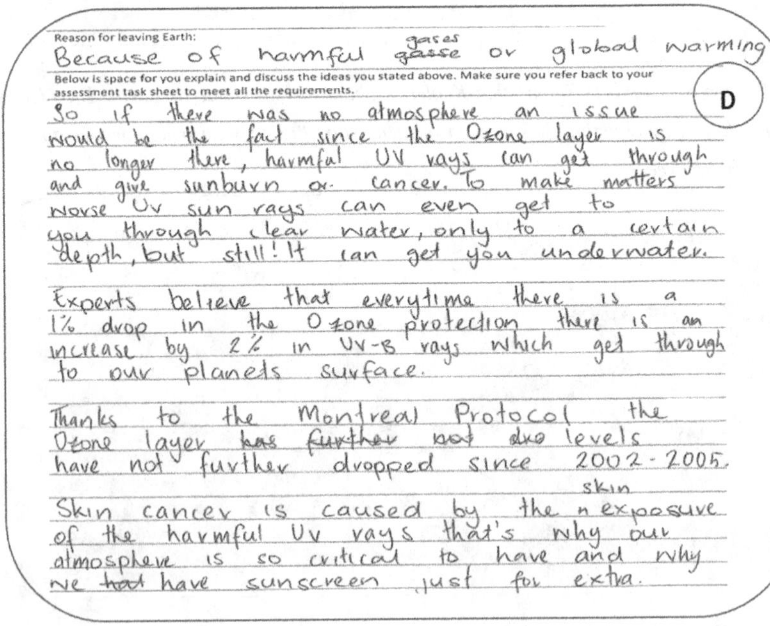

Reason for leaving Earth:
Because of harmful gases or global warming

Below is space for you explain and discuss the ideas you stated above. Make sure you refer back to your
assessment task sheet to meet all the requirements.

So if there was no atmosphere an issue
would be the fact since the Ozone layer is
no longer there, harmful UV rays can get through
and give sunburn or cancer. To make matters
worse Uv sun rays can even get to
you through clear water, only to a certain
depth, but still! It can get you underwater.

Experts believe that everytime there is a
1% drop in the Ozone protection there is an
increase by 2% in UV-8 rays which get through
to our planets surface.

Thanks to the Montreal Protocol the
Ozone layer has further not the levels
have not further dropped since 2002-2005.

Skin cancer is caused by the n exposure
of the harmful Uv rays that's why our
atmosphere is so critical to have and why
we had have sunscreen just for extra.

D

Fig. 3.9 (continued)

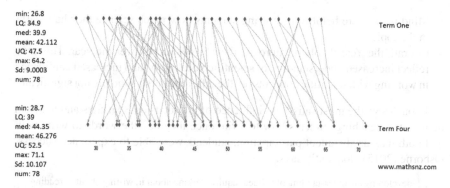

min: 26.8
LQ: 34.9
med: 39.9
mean: 42.112
UQ: 47.5
max: 64.2
Sd: 9.0003
num: 78

Term One

min: 28.7
LQ: 39
med: 44.35
mean: 46.276
UQ: 52.5
max: 71.1
Sd: 10.107
num: 78

Term Four

www.mathsnz.com

Fig. 3.10 Results of thinking with evidence assessment

Science Thinking with Evidence Assessment

Year 9 students sat the New Zealand Council for Educational Research STwE, both in February and then again in November. The results are presented in Fig. 3.10.

Summary: (n = 78).
Improved Score60 students (77%).
No change2 students (3%).
Decreased Score16 students (20%).
Term One (Feb): Average Scale Score: 42.1
Term Four (Nov): Average Scale Score: 46.3

As can be seen there is very little change between the initial and final STwE. We decided to include this test because it was the normal practice at this school. The test presents information to the students in different forms, scenarios, graphs, other forms of text and they are required to select the most appropriate option from multiple choices offered. As we have said the central issue in the project is that of literacy and in our view most questions were inaccessible to a large number of these students. We discuss these in some detail later.

Findings

- Students in class engagement with the Sparklers supported this vocabulary development.
- Students learnt the keywords to explain the science ideas they were learning.
- There was improvement in students' ability to communicate their science ideas during discussion and in writing.
- Students experienced success and this led to greater engagement in using the Sparklers as the year progressed.

- Mind maps were helpful in showing students how much science they had learnt in the topic.
- Overall, the Year 9 cohort improved their STwE scores over the year. This may reflect increased understanding of scientific vocabulary, and increased confidence in working with scientific concepts but the difference is statistically not significant.

In our view, the most satisfying influence of the focus on Sparklers and teachers deliberately teaching students how to write once they had the words to write with had both increased student participation, students were able to experience success. Osborne (2015) eloquently says:

> .... developing an understanding of an idea requires talking about it, writing about it, reading about it and representing/drawing or visualising it (p.18)

Talking, writing, and reading about it, our students are on their way. Some may consider the Sparklers to be a collection on worksheets, which they are. However, it is an approach that is accessible to all teachers anywhere. The simplicity of the intervention is its most attractive appeal.

Reference

Osborne, J. (2015). Practical work in science: Misunderstood and badly used? *School Science Review, 96,* 357.

Chapter 4
Year 12 Case Study, Concept Mapping, and Communicating Science Ideas

Abstract This chapter presents the Year 12 case study. This inquiry involved three teachers from the Year 9 case, one each for chemistry, biology, and physics, inquiring into their teaching practice with a view to helping students to develop their literacy skills, conceptual understanding, and ability to communicate their ideas in writing. Participating teachers agreed to use concept mapping to support their students to connect the science ideas they were learning and then to scaffold students to use their concept maps to write coherent paragraphs to explain their developing science ideas.

Keywords Concept maps · Scaffolding communicating in science · Evidence of student learning

Concept maps have been widely used for formative assessment (Bennett, 2011) and have been recommended for improving science teaching and learning (Cowie & Bell, 1999). In concept maps, students present their knowledge in the form of graphical representations where students use key ideas as nodes and links showing the connectedness between the ideas (Stephenson, Hartmeyer, & Bentsen, 2017). Concept maps are interpreted in relation to what they expose in terms of what they tell about students' knowledge and understandings (Black & Wiliam, 2009). Analysis is done either through comparing student maps with a criterion map or by scoring the number of correct responses. Stephenson et al. (2017) recommend five elements for successfully use of concept mapping. These include concept mapping should be taught, and perhaps several times; if the purpose is to elicit individuals' understanding then they should be created individually; if used for formative assessment, low teacher-directed mapping is most effective; technology assisted and peer assessment strategies can be useful; feedback should be provided; and there should be student ownership.

Methodology

The methodology used with Year 12 was pre- and post-concept maps and cogenerative dialogue (Cogens) (Tobin, 2006). The Cogens are regular but informal meetings where small groups of students and teachers meet to talk about learning and how it can be improved. The teacher approach is 'how can I best help you to learn?' Tobin promotes the effectiveness of engaging students and of teachers talking about learning and how it can be enhanced in a safe environment where teachers are responsive to student feedback. As with the Year 9 case study, we found that students were not responsive during the first Cogen. Pasifika students are polite and find it difficult to suggest that teachers should teach differently. We replaced the Cogens with focus group interviews conducted by a former student who was well liked by the students and they shared their thoughts with.

The main strategy was developing concept maps to think about and to link science ideas learnt within a topic. The strategy had been researched and found to have positive outcomes for students in a high decile New Zealand school (Moeed & Easterbrook, 2012). Teachers helped students to develop their writing skills. The intention was to support students to improve their science vocabulary and learn to communicate their developing understandings of science ideas. One concern the teachers had was that their students performed poorly in assessments as they have lower literacy and were not confident writers.

The Inquiry Process

See Fig. 4.1.

Year 12 Results

Concept maps were the strategy used to help students learn in Physics, Chemistry, and Biology classes. The teachers found that the skills for constructing a concept map had to be taught and practiced, which should not have been a surprise as they had not been taught it. In most classes, the teachers helped students to construct their initial concept maps. Time was allowed in class for students to link their new ideas to their existing concept maps. During the year they became increasingly skillful at drawing these maps and reported that they were useful. Other learning strategies were encouraged and included revision, constructing sentences to communicate ideas, and paragraph writing.

> **Focus of Year 12 inquiry**
>
> To teach concept mapping to Year 12 students and support them to make links
> between ideas within a science topic

> **Teaching Inquiry**
>
> 1. Help students to draw a concept map at the start of a topic.
> 2. For three lessons each week teach the topic and help students to learn the key ideas.
> 3. In the fourth lesson each week give students 10-15 minutes
> to add ideas learnt to the original concept.
> 4. Check maps for accuracy.

> **AKO**
>
> (Māori word for reciprocity of teaching and learning)
> Encourage and support students to participate in this learning activity.

> **At the end of the topic**
>
> 1. Teachers check student's concept maps.
> 2. Survey students' views about the usefulness of concept maps for
> learning.

> **Research team evidence-based evaluation of the strategy**
>
> 1. Collaboratively consider what has worked well and what may need to change.
> 2. How robust was the evidence? What may be a better way of measuring effectiveness?

> **Make appropriate changes**
>
> **Repeat the inquiry with the next topic**

Fig. 4.1 The process of Year 12 inquiry

Results of Year 12 Student Interviews

It appears that students were initially reluctant to talk about what helped and did not help them to learn and their responses were brief, but they were happy to provide more detail in the second interview, perhaps because they began to trust the interviewer or they were becoming more confident with the learning strategies.

Students were interviewed about their learning in their Physics, Chemistry, and Biology classes. The interview transcripts were coded, and what students identified as helpful is summarised in Fig. 4.2. All names used are pseudonyms.

Students said that they constructed concept maps in all their Year 12 Science subjects and found them useful. Initially, they often tended to draw mind maps or brainstorms rather than concept maps. Early in the process one student said:

> I kind of like them. It's pretty helpful but I'm just not good at putting the words on the lines. But I can just put all the information around ... that connects to the main idea. ... Like I just don't know how to link it to the other ideas. (Sam, Chemistry student)

Several students said that the concept maps helped them to construct sentences and paragraphs:

> ....because you can put words on the lines, so it goes from one idea to another. If you go from that one idea to another you can read across the line and it would make like a sentence. (Vai, a Biology student).

> So if we have to write a whole paragraph for something I can go back to the concept map and have the first five words of the paragraph and then I'll know the whole thing...It just makes sense. Mr…, showed us how to build paragraphs from a concept map. (Soa, a Physics student).

> We wrote a paragraph and we had to use an equation to try and explain it. Yeah, the concept map actually helped me write sentences. Because we did a before paragraph, but I didn't write much...and [then we did] this one using the concept map. We were able to write a more detailed paragraph. (Tasi, a Chemistry student).

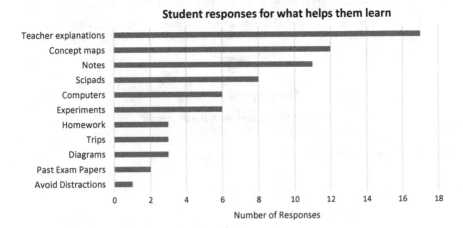

Fig. 4.2 Students' views about what helped them to learn

Some students liked the concept maps breaking the topic down into small pieces, for example:

> So the mind maps oh concept maps, they do break it up and yeah it makes it easier...I think it's because when I draw them...and then I draw the lines and it like just connects and then I can come back to it and I can just understand, and like because the mind maps they break up the sentences that we have to write. (Maia, a Biology student).

Some confusion about mind maps and concept maps. Students also appreciated that concept maps provided an overview:

> It helps link the words together. We just come up with any idea and then yeah...Just throw out the ideas and just link them together, it is cool to do concept maps. (Nunu, a Chemistry student).

> Because it helped me see all the ideas that could come with it. I think this is cool, because it lets me see how it all connects as well. (Grace, a Biology student).

Some students commented on the opportunity concept maps provided for feedback and improvement:

> Sometimes what I write on the lines doesn't make sense. Yeah it doesn't make sense in my head...I'll read over it and think of what would actually go there...And then re-write it. (Tolo, a Physics student).

> These maps helped me put down what I know and go through what I don't know, I can improve. And if I don't like a point, I can improve it. I like how you can be really specific with them.... I put down what I know and Miss.... looks through it and says, 'you can improve this here'. (Tasi, a biology student).

Some Physics students found concept maps useful for revision. For example:

> Concept maps help me recap what we have learnt and expand it a bit more and like just make it stick in our brain. (Mere, a Physics student).

> Like when I'm revising it helps because all my information is just there. I just have to find it myself. So, we have exams coming up. Like that's kind of my way of studying. I like it. I draw them at home too...Yeah, because I like visuals. (Fatai, a Physics student).

Tolo, a Chemistry student made an insightful comment. He said that in Chemistry he finds flow charts more useful.... "Like one thing follows another". But when he is making concept maps in Biology "I can see how ideas connect with each other".

Results of Student Concept Maps

Initially, students often tended to draw mind maps, or brainstorm rather than concept maps. When they began to draw concept maps there were often few words linking the concept boxes (see Fig. 4.3) or confusion between concepts and linking words, with both being put into concept boxes (see Fig. 4.4).

Initially, students' concept maps at the beginning of a topic contained only a few concepts and were relatively simple. As the topic progressed, students were able to

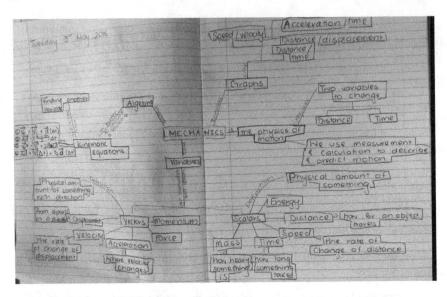

Fig. 4.3 Concept map with few linking words

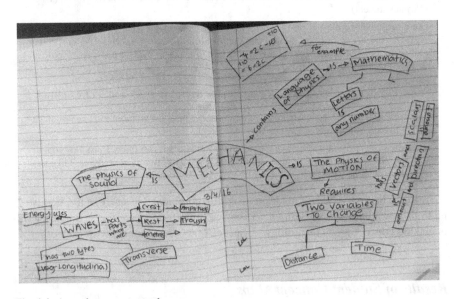

Fig. 4.4 An early concept map 1

add to and develop their concept map. Figures 4.3, 4.4, 4.5 and 4.6 show concept maps at different stages of a physics topic.

Figures 4.7 and 4.8 show concept maps drawn by a chemistry student. The initial maps 4.7 shows few concepts.

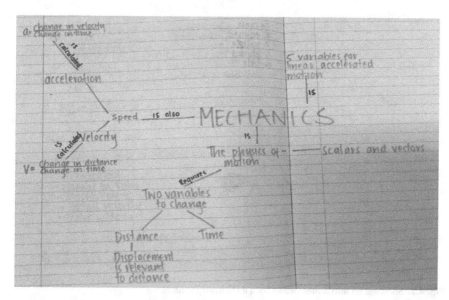

Fig. 4.5 Another early concept map

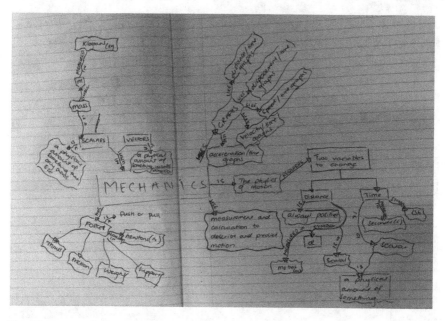

Fig. 4.6 A more advance concept map

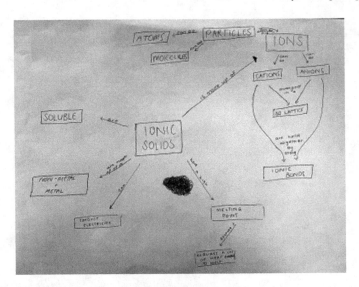

Fig. 4.7 Showing an initial map drawn by a a chemistry student

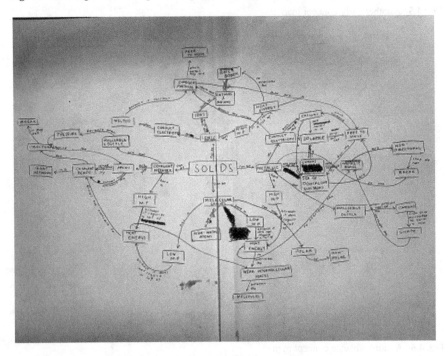

Fig. 4.8 Showing a final map drawn by the same chemistry student

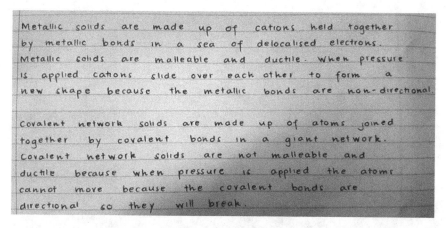

Metallic solids are made up of cations held together by metallic bonds in a sea of delocalised electrons. Metallic solids are malleable and ductile. When pressure is applied cations slide over each other to form a new shape because the metallic bonds are non-directional.

Covalent network solids are made up of atoms joined together by covalent bonds in a giant network. Covalent network solids are not malleable and ductile because when pressure is applied the atoms cannot move because the covalent bonds are directional so they will break.

Fig. 4.9 Explanatory paragraphs developed using a concept map

However, in Fig. 4.8 (advanced concept map) many concepts (boxes) are added, many new links made, and large number of linking words and phrases placed on the arrows. Unfortunately, the image in not clear. However, the teacher reported that the student had added several boxes and made correct links.

Writing Paragraphs

One of the challenges that the teachers had identified was that their Year 12 students did not have the skills to write paragraphs, one teacher saying they did not have the words to construct sentences, let alone write paragraphs.

The advanced concept map in Fig. 4.6 was then used by its author to construct paragraphs explaining malleability and ductility in different types of solids, contrasting metals with covalent network solids. These paragraphs can be seen in Fig. 4.9. This student obtained a Merit grade in the external examination for this topic.

In Physics, Jemma created a concept map on the force's topic (See Fig. 4.10). This she then used to write a paragraph shown in Fig. 4.11.

Aroha who had helped Jemma in constructing her concept map, wrote the paragraph shown in Fig. 4.12. There are similarities between the two student's paragraph writing as they had been working collaboratively.

Mock Exam Results

The mock exams were the first assessment after having worked on the concept maps. The number of students was 15 (Tables 4.1).

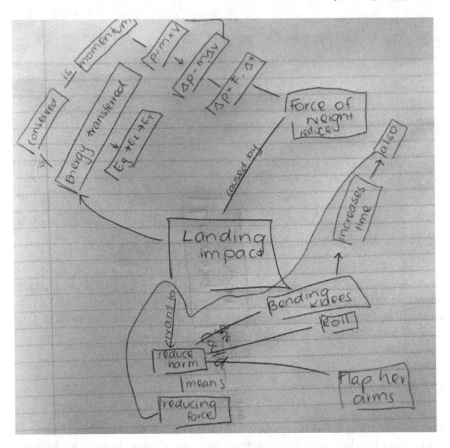

Fig. 4.10 Jemma's concept map

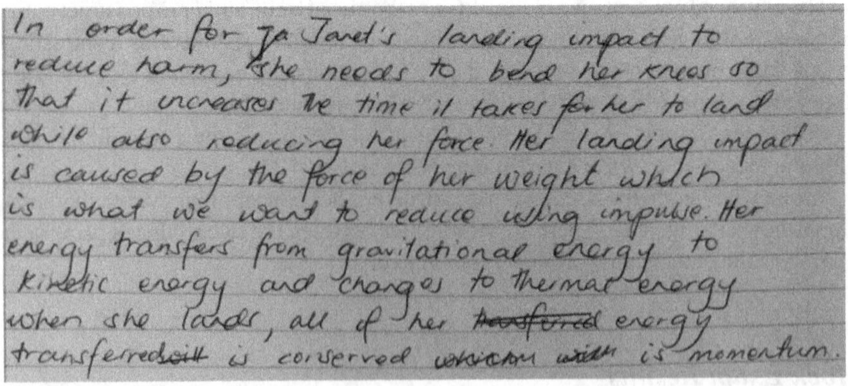

In order for ~~Ja~~ Janet's landing impact to reduce harm, she needs to bend her knees so that it increases the time it takes for her to land while also reducing her force. Her landing impact is caused by the force of her weight which is what we want to reduce using impulse. Her energy transfers from gravitational energy to kinetic energy and changes to thermal energy when she lands, all of her ~~transferred~~ energy transferred~~it~~ is conserved ~~which~~ ~~with~~ is momentum.

Fig. 4.11 Jemma's paragraph developed from the concept map shown in Fig. 4.10

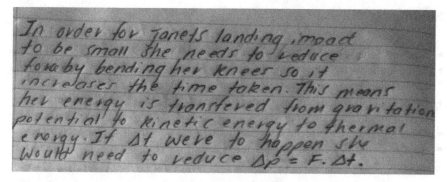

In order for Janets landing impact to be small she needs to reduce force by bending her knees so it increases the time taken. This means her energy is transferred from gravitation potential to kinetic energy to thermal energy. If Δt were to happen she would need to reduce Δp = F.Δt.

Fig. 4.12 Aroha's paragraph developed from the concept map shown in Fig. 4.10

Table 4.1 Results of the mock examination

Not achieved	Achieved	Merit	Excellence
7	5	3	0

National Certificate in Educational Achievement (NCEA) Results

In New Zealand National Certificate in Educational Achievement for qualification are offered as NCEA Level 1, 2, and 3. The participating Year 12 students are assessed for NCEA Level 2 qualifications. The assessment is through Achievement Standards. Some Achievement Standards are internally assessed at school whereas others are externally assessed.

The Table 4.2 summarises the results of external Level 2 Science examinations for the research school from 2012 to 2016.

Results for Achievement Standards 91,157, 91,164, 91,165, 91,167.

Table 4.2 shows an improvement in the proportion of students gaining an Achieved grade or better in 2016, compared to earlier years. There has also been an increase in the proportion of students gaining Merit and Excellence grades.

Table 4.2 Results of Level 2 Science examinations from 2012–2016

Year	Percentage				Actual numbers				
	Not ach (%)	Achieved (%)	Merit (%)	Excellence (%)	Not ach	Achieved	Merit	Excellence	Total
2012	89	11	0	0	32	4	0	0	36
2013	90	10	0	0	26	3	0	0	29
2014	79	14	7	0	11	2	1	0	14
2015	82	16	2	0	42	8	1	0	51
2016	63	25	9	3	20	8	3	1	32

Findings

- Students found concept maps supported their learning
- Some students found concept maps to be a means of identifying gaps in their knowledge
- Some students found concept maps provided a medium for seeking feedback and improving their understanding
- Other students were able to construct written answers using their concept maps
- Students said they found concept maps useful for recapping their ideas and for exam revision.

The teachers kept evidence of student work and this was shared at the hui after the first cycle and the second cycle. Students became increasingly confident in both drawing the concept maps as well as writing paragraphs. The next Chapter reports the Teaching as inquiry which all the teachers conducted.

Chapter 5
Teaching as Inquiry

> *The Success or Failure of My Students is About What I Do. I Am a Change Agent.*
>
> John Hattie (2013)

Abstract Teaching as inquiry is research undertaken by teachers to improve their practice. In this chapter we present the findings of our research where the participating teachers inquired into their science teaching practice. We report the findings of their inquiries. A key point about this statement is that a pedagogical approach that may have a positive outcome on student learning in one school may not have the same impact in another, essentially, context matters. *The New Zealand Curriculum* (Ministry of Education, 2007) encourages teachers to use this as a tool for improving their teaching with the focus on better outcomes for all students and states that, "Since any teaching strategy works differently in different contexts for different students, effective pedagogy requires that teachers inquire into the impact of their teaching on their students" (p. 35).

Keywords Teaching as inquiry · Evidence-based decision making

At the first research group meeting with participating teachers, we started by considering the following questions proposed by Sinnema and Aitken (2015, p. 134) (Fig. 5.1).

A thoughtful discussion followed, and we collectively concluded that Teacher inquiry into our practice would be a good idea. The concerns raised included, students not engaging, not being able to write, not bringing their books to class, and not coming to class and less than optimum achievement in science. Teachers talked about what they were currently doing that was effective for one class but may not for another.

Monitoring is a means of sustaining the progress being made by students and the teachers in this research were aware that their students were not making the progress the teachers intended the students to make. What might be the reasons for poor achievement? One suggestion was that for assessment in senior school, students needed to be able to write paragraphs, which they were unable to do. That told us that literacy was perhaps one of the biggest issues. To be able to communicate science

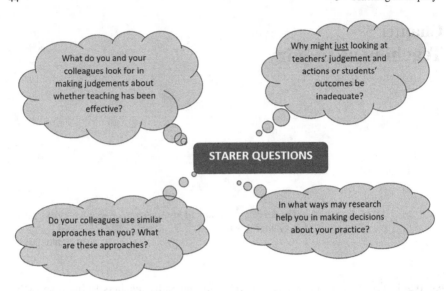

Fig. 5.1 Starter questions used to start the conversation

ideas students needed language proficiency. During further discussion it became clear that if teachers wanted their students to succeed in senior school, teachers needed to help students to improve their literacy in junior school. Literacy alone was not going to be helpful either. Students needed to be familiar with the science vocabulary, understand what it means, and then communicate it either in writing, verbally or through representations of some form.

Teacher were provided with a research article reporting the findings of research carried out in another nearby school (Moeed & Easterbrook, 2014). That research reported that the repeated use of Sparklers, a collection of worksheets written specifically for helping students to learn science vocabulary had been useful for increasing engagement in science classes. In senior school, concept maps had been an effective way for students to organise their developing ideas. Conner (2015) in her New Zealand research into priority learners found that when teachers experienced positive outcomes during an inquiry cycle, it motivated them to continue with the next iteration of changes to their practice. Māori students and students from other Pacific nation backgrounds were identified as priority learners by the Ministry of Education at the time. Most students in Conner's school were from these groups, therefore priority learners. Conner noted that support and time were the key factors for enhancing teacher agency to develop their practice. The Ministry of Education initiative "Teacher-led Innovation Fund" would give the teachers this support and time.

The research presented here involved all four teachers Sue, Bev, Tim, and James who did the inquiry in their Year 9 classes and one physics, one chemistry and one biology teacher in their Year 12 classes. First a pilot study took place to find out if

Fig. 5.2 Initial planning for Year 9 inquiry

the two chosen strategies would be effective. A package of Sparklers was bought from Shelley Monds, a local teacher and writer of this resource.

Year 9 Teacher Inquiry

How Year 9 inquiry was conducted, and its findings are reported in Chap. 3 and represented in Fig. 5.2. Similarly, initial planning of Year 12 inquiry is shown in Fig. 5.3 and the findings are in Chap. 4.

Year 12 Teacher Inquiry

After the pilot phase, the first cycle of the project started. Teachers used *Sparklers* as a starter activity with their classes. After completing the first topic of inquiry with Year 9, a full team reflective meeting was held, which included the four participating teachers, the school management representative (Deputy Principal), and the project advisor. The conversation was audiotaped and transcribed. In this reflective session the discussion was generally positive about the use of Sparklers with the teachers making the following comments.

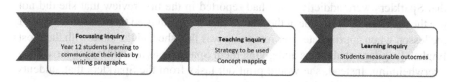

Fig. 5.3 Initial planning for Year 12 inquiry

Change in Teacher Practice in the First Cycle

- Sparklers were a useful routine and got the students in the habit of settling in (Three teachers used it as a routine, and one teacher used Sparklers only when it fitted with her planned teaching, but she will consider establishing it as a routine).
- Most students were engaged although some needed support.
- There has been a noticeable gain in student science vocabulary.
- Students were more willing to participate in discussion as they *had the science words* to confidently communicate.
- There are extra challenges to the ESOL (English for Speakers of Other Languages) class doing the Sparklers, but the teacher was able to support students to do so.
- There was some concern that students were looking for words to complete the task rather than reading all the material. It was agreed that in the very first go, for some students to just be able to look for words was progress. Upon discussion, it was agreed that a second way and more meaningful learning of the science words could be to do a sentence writing task to go with the Sparklers where students are given three/four words they have been learning and asked to write a meaningful sentence.
- A second strategy to be included was to do a test with ten questions to see if the keywords are understood. The questions are quiz type questions and would allow measurement of student progress during the topic.
- It was agreed that it was a good start and for some students it will take longer to take on board new learning routines but with support this appeared to be achievable.

Similar reflective meetings took place at the end of each inquiry cycle and agreed changes were subsequently implemented. For example, in the second reflective session one teacher noted that, "my students are keen to help each other" and asked is that ok? (Sue) Another teacher who was unsure if student's interest in the Sparklers would continue reported, "They come in, go straight to the Sparklers, and have their heads down until asked to stop, I wonder if this will continue" (Tim). Bev agreed that Sparklers were addictive. Sue had reported in the first review that she did not use the Sparklers as a routine at the start of the lesson but used them at some stage of the lesson when appropriate. Her reasoning was that she had the class with the most English language challenges and giving them Sparklers as a booklet may have been overwhelming. Instead, she would print out tasks from the Sparklers that students could complete and post into their books.

Change in Teacher Practice in the Second Cycle

Reflection on teaching and learning has resulted in several changes in practice.
   Since the last checkpoint:

- The Sparklers are now being used in a more targeted way to support vocabulary

- Each Year 9 unit of work now has a set of topic related keywords
- Keywords are supported by the Sparklers
- Keywords provide a focus for planning
- Keywords help track student progress
- An online vocab quiz has been developed using google forms
- Summative tests have been redesigned
- A strategy has been developed to scaffold students into talking about their learning

Year 12 Study Has Begun

- Teachers have developed a strategy for teaching students how to create concept maps
- Teachers create their own concept map before teaching a Year12 unit

One noticeable change observed by the researcher was that teachers were more confident. "They were not following *instructions* or a *process*, having gained confidence they were bringing their expertise, knowledge of their students, and were making evidence-based judgements about student's learning" (Researcher observation, Term 2 of the project).

Teachers' Sense Making Conversations

Sue: I took a different approach and we started with a couple of video clips about the solar and lunar eclipse process and then we read an article that had some explanation about how it happens. So, their focus initially was to get one of the main ideas because we have been doing a bit of work on how you identify keywords and main ideas. So as a class we then put up our main ideas on the board and then we talked about linking words in the class. We were able to come up with a lot of linking words off their own back which we put up on another part of the board. And then they looked at the main ideas had come up with as a class.

Bev: Was this with the keywords sheets or without?

Sue: No, it was just on the board with the main ideas.

Bev: Yeah, ok.

Sue: And then they took those main ideas and those linking words or any other linking words they wanted to use, and they wrote their ideas about how the solar and lunar eclipses occurred. They were making sure that what they were writing down linked together and made sense. For some, it was effective, for others it was not so effective. They all could link two or three ideas together. Quite often it was just by writing three sentences and putting an 'and' or a 'therefore' in between. The following sentences was constructed collectively by the class (Figs. 5.4 and 5.5).

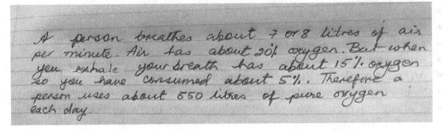

Fig. 5.4 Sue' class linking together sentences

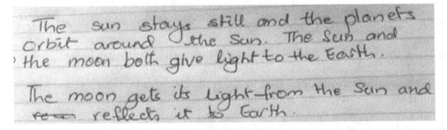

Fig. 5.5 This was written by Student B after the above collective example

Sue: I was also impressed with their ability to come up with a list of linking words because we had only really done one other activity where we had the table with linking words on it. You could see the sense of satisfaction they got from this effort.

Tim: With my class we've spent two or three lessons spread out doing bits and pieces towards paragraph writing and looking at different ideas that they've actually got something tangible that they can link together. And I think that they're quite good at seeing how ideas connect to one another and they needed a lot of prompting and scaffolding for them to actually get to a point where they felt comfortable to write down their ideas which made it a bit harder. Scaffolding suggestions, for example, "but perhaps it could have been"… "or should have been"… They felt comfortable linking their ideas after that prompting and scaffolding. From what I've read they've done quite well as a class. Most students completed the class writing three paragraphs using the small subset of linking words I gave them (Fig. 5.6).

Bev: But is some of your class still at the point where constructing a sentence in English…is a challenge?

Tim: Yes, but the REAL difference is that they are not finding every excuse under the Sun, to avoid writing".

The above, self-explanatory conversation illustrates that progress was being made. James who had been listening to the conversation added the bonus is that they are so keen on the Sparklers that they get in the class and get on with them. I have noticed, they don't need to go to the toilet as often… (laughter). A lot of collaboration is also happening.

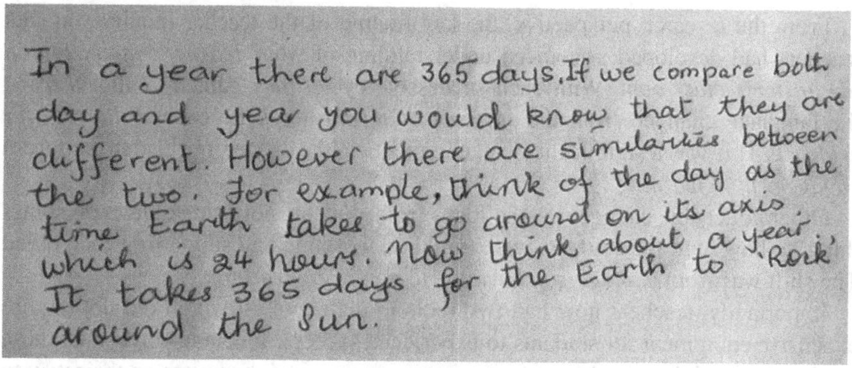

In a year there are 365 days. If we compare both day and year you would know that they are different. However there are similarities between the two. For example, think of the day as the time Earth takes to go around on its axis, which is 24 hours. Now think about a year. It takes 365 days for the Earth to 'Rock' around the Sun.

Fig. 5.6 Student C, a more advanced writing example from Tim's class

Researcher: How are the Year 12 concept maps going? Have you started on those?

James: I think the Year 12 students are also doing well. The big surprise I had was that they had a misconception about concept maps.

Bev: Mine thought it was a brainstorm. We got the brainstorm up on the whiteboard. Then sorted similar ideas into groups. Finally, we thought about what main idea each of these groups was about. We did this with the metal's topic. Students called out words and John put them on the board. I gave them a few minutes to think about which ideas were similar. For example, mercury, gold, silver, and aluminium are all metals, so we agreed that we could label these as types of metals. We discussed that most metals were solid, one student said mercury is not. So that is how we got the next link (see Fig. 5.7).

A   B

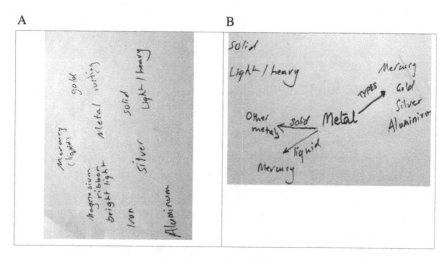

Fig. 5.7 Showing student brainstorm in Fig. A and beginnings of a concept map in Fig. B

From the research perspective, the key finding of the teacher inquiry was that teachers had developed a nuanced understanding of what *Teacher inquiry looked like in their classroom*. With each successive cycle, they gained confidence and the language changed from the *students doing the work* to teachers *looking for evidence*. This was a significant change in practice, the discussions had moved from *engagement to learning*.

Other noticeable change was the focus from students not being able to do something, to how hard the students were thinking, working, and supporting each other. The shift was from *managing behaviours* to *managing learning*.

Importantly, teachers now had two tools in their tool kit which did not require expensive equipment for students to experience success. It is noteworthy that these teachers wanted their students to experience success, and they were on their way to achieving this.

References

Hattie. J. (2013). Know thy impact; teaching, learning and leading in conversation. *Spring,1*(2) (2013). ISSN 1922–2394. Ontario Ministry of Education.

Moeed, A., & Easterbrook, M. (2014). Science teacher development through partnership with a mentor: Eight years of teaching and learning. *International Journal of Professional Development, 3*(1), 3–16.

Sinnema, C., & Aitken, G. V. (2015). Teaching as inquiry. In: Hill, M., & Fraser, D. (Eds.), *The Professional Practice of Teaching in New Zealand* (5th ed., pp. 79–97). Cengage. Retrieved from http://www.cengagebrain.co.nz/shop/isbn/9780170350716

Chapter 6
Discussion, Conclusion, and Final Thoughts

Abstract In this chapter we discuss the results of our research supported by relevant extant literature. We share the findings and the implications this research has for teacher practice and future research.

Keyword Disciplinary literacy is needed to engage and achieve in science

Success is motivational but for various reasons the participating students were not experiencing the success that the teachers had intended them to. These participating science teachers tried different strategies with varying degrees of success as these teachers were aware that to succeed in science, students needed to be able to communicate their ideas. Therefore, these teachers set out to inquire into their teaching practice to learn the process of teaching as inquiry to improve student learning and thus to enhance student achievement. Learning in science includes doing, reading, writing, talking, visualising, and representing in science (Osborne 2015). Learning from doing also needs to be communicated. Participating students brought rich funds of knowledge and culture to the school, but they generally had lower English literacy. To succeed in school science, students needed to learn scientific words and their connectedness within the discipline as there was a need to be able to understand and link science ideas in a meaningful way.

Emergent Themes

Five themes emerged in this research project; these were:

- Disciplinary literacy is needed to engage and achieve in science.
- Simple and accessible tools can lead to powerful learning.
- Concept maps are an excellent learning strategy rooted in constructivist theory of learning.
- To practice Teaching as Inquiry, teachers need a nuanced understanding of the process.
- Teacher practice can change, and this change can be sustained over time.

We discuss these themes in relation to the extant literature.

- **Disciplinary literacy is needed for EAL students to engage and achieve in science.**

  Frankel et al. (2016) define literacy as the process of using reading, writing, and oral language to extract, construct, integrate, and critique meaning through inter-action and involvement with multimodal texts in the context of socially situated practices. They emphasise that literacy involves two kinds of processes, produc-tive processes (e.g., writing and reading) and receptive processes (e.g., reading and listening). In this study, not having the words to talk with was recognized as the problem teachers wanted to solve. In science learning, literacy is fundamental in both hands-on experiences and to understand others' investigations. There is a synergy between science and literacy as both these areas aim to foster similar skills. For example, science requires students to "make sense of data, draw infer-ences, construct arguments based on evidence, infer word meanings, and of course, construct meanings for text—the very dispositions required as good readers and writers" (Pearson et al., 2010, p. 460). We found that once Year 9 students had the topic specific vocabulary; they were more engaged. Further, they were willing to try and write down their science ideas. The collaborative approach taken by some teachers in the beginning worked well for those students that needed extra support.

  The decision to begin with Year 9, the first year at the participating school, meant that the strategies that the teachers used with confidence by the end of the project helped an entire cohort of students to progress to the next level with better science literacy. We have seen evidence of improvement in what Beeby and Roberts (2014) identify as students' ability to apply skills and knowledge. With reference to Bybee's (1995) dimensions of literacy, we have presented evidence of students having developed *functional literacy*, having a bank of science vocabulary to draw upon and to begin communicating their science ideas both verbally and in writing. They were also developing *conceptual literacy* which became evident in improved test marks. The focus of the forensics unit was on *procedural literacy* and although the students had familiarity of the investigative procedures, there was little evidence of their being able to use these in different contexts, so it was limited.

  Osborne (2002) argues that, "science is more than its vocabulary; words have value only when used as referents or to represent meanings" (p. 212). In this research, evidence suggests that students had made considerable progress in vocabulary acquisition, but they were also able to use this in the right context and to demonstrate their understanding of the science ideas.

  The results of the Science Thinking with Evidence (STwE) assessment also high-lighted two important points, one being that the students did not have the literacy or experience of taking part in this test which does not make the results cred-ible. The other important aspect was that the project did not focus on *multi-functional literacy*, which in this case would have been developing an under-standing of the nature of science, or how science works. The STwE test assesses students' understanding about how science works and the concept of evidence. Research suggests that sometimes standardised tests are not used for the purpose for which they were designed. For example, Horsley and Moeed (2018) found that the Progressive Achievement Tests (PATs), which assess students' Mathematics, Listening Comprehension, Punctuation and Grammar, Reading Comprehension, and Reading Vocabulary, were being used to identify students with high ability in science. These tests do not measure any aspect of science learning. Perhaps schools, like the one participating in this research, are using the StwE test to evaluate whether students have made progress in science learning, although the test is neither designed to measure all science learning nor is it suitable for the predominantly Pasifika student population. Additionally, judging students' multifunctional science literacy through tests they perhaps cannot comprehend is unlikely to be useful. The assessment in this case is not valid and its use does not pay attention to the tenet that assessment should do *no* harm (Moeed & Hall, 2011). Year after year to be told that the students have made little progress is highly unlikely to be motivational for either the students or their teachers.

This brings us to another aspect of motivation that had not been the focus of this research but nevertheless is an important finding. All teachers in the inquiry at different stages talked about students enjoying working with Sparklers and the various ways in which Sparklers resulted in higher student engagement; the students liked them because they experienced success, they could do them, and that they were addictive. The Sparklers, which are a collection of worksheets, were attractive although the teachers said that students did not enjoy doing worksheets and consequently, they were left behind; if worksheets were set for homework they did not come back completed. The teachers wondered though whether easier tasks motivated students to *want to learn* because they were experiencing success or whether having challenging tasks to start with would give students the same sense of achievement (Midgely, 2014). Palmer (2009) suggests that motivation is both a pre-requisite and co-requisite for learning. Perhaps engagement in the Sparklers, and their success, is being able to complete the tasks that have a positive influence on learning. Corno (1992) argues that motivation to learn requires volition to carry out the learning and that volition has two elements, "the strength of will" needed to complete the task and the "diligence of pursuit" (p. 14). Corno and Kanfer (1993) argue that motivational processes only lead to the decision to act. Once the individual engages in action, volitional processes take over. We cannot say that initially there was strength of *will,* but the students certainly had the diligence of pursuit when it came to Sparklers.

Researchers for example, Cervetti et.al. (2012) and Goldsmith (2010) recommend that literacy learning strategies woven throughout lessons and combined with hands-on investigations had positive outcomes for students and heightened student motivation. They found that test scores improved on reading comprehension, science, and topic specific vocabulary which is similar to the findings of the present research.

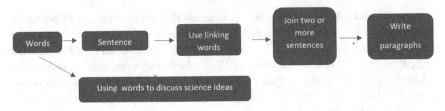

Fig. 6.1 Scaffolded progression from words to paragraph writing

- **Simple and accessible tools can lead to powerful learning**
  One can look upon the Sparklers as a collection of worksheets, but that would not be fair. They are designed specifically for increasing vocabulary and understanding of the related science ideas (see an example of a Sparkler in Appendix 1). The structure allows for engaging with the set vocabulary in multiple ways. As students work through the Sparklers the teachers are also teaching those same science ideas either through practical work or engaging students in other learning activities.
  We think that structure is very important. Making the Sparklers as a routine had the added benefit that students settled in quickly and were focused on learning. Providing support through the linking words sheets (See Appendix was beneficial in at least one of the classes (Fig. 6.1).

- **Concept maps are an excellent learning strategy rooted in constructivist theory of learning**
  The constructivist theory of learning states that students learn new ideas in relation to what they already know (Driver et al. (1994); Driver, & Bell, 1986). It is therefore important in constructivist pedagogy to first find out what the learner already knows and then identify any misconceptions (Baviskar et al., 2008). From a social constructivist perspective knowledge is individually constructed but socially mediated (Leeds-Hurwitz, 2009; Solomon, 1995).
  Concept maps have been used in for meaningful science teaching and learning. They are a technique that visually illustrates the relationship between science ideas. Students have used concept maps as a study tool and teachers have used them to evaluate student's understanding of science (Akay, 2017).
  In their meta- analysis of studying and constructing concept maps, Schroeder, Nesbit, Anguiano and Adesope (2017) state that scholars cite various reasons for the effectiveness of concept maps as an effective learning strategy and these reasons include, "promoting meaningful learning, reducing extraneous cognitive load, or both" (p. 433) These reasons appear to be true for the participants in our research.
  Students found concept maps helpful, learnt how to draw and use them for sorting out science concepts and the relationships between them. Additionally, the process encouraged them to communicate their understandings of the science ideas both orally and in writing. Since all assessment in senior school, even practical investigations are assessed through a pen and paper assessment using this strategy to enable students to communicate in writing is a useful finding.

- **To practice Teaching as Inquiry, teachers need a nuanced understanding of what Teaching as Inquiry is**

  Teacher learning and how they acquire new knowledge and ideas, perhaps changing or even deleting old practices is a key ingredient in any educational reform. Some researchers propose that just like student learning, teacher learning ought to be underpinned by constructivist theory of learning (Feldman, 2000; Borko & Putnam, 1996). Feldman (2000) explains that just like a learner must become dissatisfied with their learning and with their understanding, a teacher must also become discontent with their practice because they realise that it is, "ineffective, unsuccessful, or because it leads to dissonance or dilemmas in practice"? (p. 612). Further, the dilemmas are due to the teacher's belief that something in morally, ethically, or even politically wrong in their practice.

  Such a situation arose for the teachers of the science department in the research school. The political focus was on raising the achievement of Māori and Pasifika students in science. Teachers had individually tried approaches they had heard about or found out on a one-day professional development course. They had concluded that what they were individually trying was not working for their students. Knowing that they can access resources that will give them some time to inquire into their practice was welcomed by them.

  They were presented with strategies that had worked in a large school in the same city and research had found these strategies to be effective (Moeed & Easterbrook, 2016). Teachers were willing to try these two strategies: Sparklers, and concept mapping. They had the ownership of the project and all decisions were made collectively. They had regular meeting times to come together, look at the data collected, share classroom experiences, even if they were things like, "For the first time Jonny had a go at starting a concept map and was willing to get help for his friend" made them a team, sharing a journey together and experiencing sometimes, small successes. The destination looked closer with each successive cycle.

  As the findings presented in this book illustrate, teachers changed their practice with a total focus on student learning and outcomes and they sustained this through successive cycles. Teachers knew how to do "Teaching as Inquiry" as required by the curriculum, they had inquired, and they had changed their practice and had experienced their students engaging and learning.

  In inquiring into teaching practice, the teachers needed the evidence of learning and progression (Talbot-Smith et al., 2013). The research design allowed for the collection of evidence of student learning using a variety of tools, both formal and informal. The teachers found weekly research team meetings, although time consuming, were very useful to try to make sense of the evidence they collected.

  Using multiple sources of evidence, and creative methodologies, and assessing in a variety of ways gave the teachers, and researchers rich insights into student learning as recommended by Rennie et al. (2003). The teachers were confronted with a mountain of data and learnt that not all students did things that they had planned for them. Teachers learnt about the messy business of doing qualitative research which was neither straightforward nor linear (Lambotte & Meunier,

2013). Together, the researchers and teachers discussed, debated, and made sense of the data that have been presented here. The teachers went through three cycles of inquiry, each time deciding on an agreed aspect to work on. In using Sparklers with the juniors in the first cycle, the difference noticed was the increase in the level of student engagement. In most classes, students came to class, and began with a task, which became a routine. They wanted to keep doing the tasks, sometimes by themselves and at other times seeking help from their friends. In our view, it created a culture of *wanting to learn*. Evidence of success in learning science vocabulary initially came in small ways, but the students began to have words to write and were able to discuss them. The teachers found that the Sparklers were a useful tool for many students, but not for all. Some students found them easy and raced through them, so the teachers made sure that there were more challenging tasks for these students to do in later cycles. The teachers wondered though whether easier tasks motivated students to want to learn because they were experiencing success or whether having challenging tasks to start with would give students the same sense of achievement (Midgely, 2014). The test results showed that students not only knew the words but were demonstrating an understanding of the associated concepts. Evidence suggests that most students improved as the units progressed. Motivation, it appears, was both a pre-requisite and co-requisite for learning as Palmer (2009) suggested.

In the Year 12 inquiry, the plan was to use concept maps, but here the teachers found that students did not have the words to start the maps and needed to be taught how to draw them. Their initial maps were like brainstorms, and they progressed to mind maps, but by the third cycle more students could draw concept maps. Teachers learnt to put more time into modelling the process of making concept maps and realized that further work was needed on this aspect. They found it useful to start students on the concept maps and keep linking their new learnings to the original concept map, very much a constructivist learning approach. The teachers expressed confidence in their ability to use the inquiry process, what evidence to gather and how to use it to support student learning.

Putting in place student learning in Year 9 meant that in the following year, Year 10 students would be starting the year not only knowing how to use this learning strategy but also, they would have rehearsed it in three successive cycles. Year 12 students had been taught how to do concept maps and use them to write paragraphs. The teachers at first reported that the students did not know how to draw concept maps and that they had to teach them. An important learning for all was that it is a skill that needs to be *taught* and *rehearsed*. There was a marked improvement in the number of ideas the students were able to link in a meaningful way. However, there was insufficient evidence in the details to do a consistent analysis. At best it can be said that there was a positive change in the way the students were representing their science ideas. Being able to write coherently had been a challenge for these students for many years, but it appears that the teachers were teaching a strategy that took students from learning science ideas to connecting those ideas and being able to communicate them. STwE test results and the NCEA results have shown an improvement, which is promising for the students progressing to Year 10 and Year 13, respectively.

Perhaps the most interesting and noticeable change was how the students came into the classroom and quickly began their work. The teachers also reported a positive difference in the attendance of the senior students in their classes.

- **Teacher practice can change, and this change can be sustained over time.**

What teachers know, believe, and their experiences have a major impact on how they teach.

Teacher's professional development that focusses on student learning and aims to enhance student learning of specific content is likely to have positive effects on teacher practice (Blank, de las Alas, & Smith, 2007). In this research, the professional development was in the form of teaching as inquiry to support students to develop their literacy and science literacy. Teachers also wanted to improve their practice so that they could make evidence-based decisions about their teaching through systematically gathering, analysing, and critiquing information about student learning. The collective long-term focus was to improve their teaching and to improve student learning. Darling-Hammond et al. (2009) argue that professional development where teachers actively engage in ways that enhance their knowledge of content, and how to teach it, gives them a sense of self-efficacy, particularly when the content is related to the curriculum and is in line with current policy. Research indicates that collaborative professional development, conducted in collegial learning environments, can facilitate school-wide change, and contribute to the development of communities of practice within schools (Darling-Hammond et al. (2009). During the research, a community of science learners and teachers came into being and in such communities, the community itself learns. They bring their thoughts, beliefs, and values and through communication and collaboration learn to take initiative, take risks, challenge old habits, and use old resources in new ways (Svendsen, 2015). The community innovates by using new resources in creative ways for supporting student learning. In the science department where this research was conducted teachers were passionate about science teaching and learning, and they came together as a community of teachers who set themselves a shared goal and worked towards achieving it. This community included their classes and we saw government policy on teacher inquiry was better understood, put into practice, and refined. It also became clear that when new policies are rolled out, just reading about them does not necessarily translate into practice. The teachers in this project had researcher support to come to a shared understanding of *teaching as inquiry*. They had the time to read relevant research, discuss and share ideas, and consider what credible evidence might look like that enabled them to inquire into their practice. We found that this research and practice partnership was a collaborative one where the teachers were involved in all stages from the planning to analysis of data, making sense of it and deciding the next steps of inquiry. Relevant to this context, Darling-Hammond (2016) asserts that teachers who succeed at developing deep understanding of challenging learning issues involving students identified as traditionally thought to be at risk, help to develop:

...student confidence, motivation, and effort and assure that students feel connected and capable in school. Their strategies for supporting learning extend beyond technical teaching techniques. They practice what John Dewey called "manner" as method: Their commitment to student learning and success supports students in the risky quest for knowledge. (p. 86)

Conclusion

Evidence suggests that participating students improved in each successive cycle. Sparklers provided the words, students did the rehearsal, and the teachers maintained the routine, provided the scaffolding necessary, and gathered evidence. Sparklers are an inexpensive resource that can be put together according to the topic and need of the class and most schools can afford it. In our view they worked because of the process followed.

The positive unintended consequence of this research project was a change in teacher focus on student learning and away from behaviour management issues. Perhaps a negative was that the cogenerative dialoguing did not work for this project as a tool because culturally, these students respected teachers and did not like to openly talk about what they were thinking and were hesitant to talk to the teachers. However, interviews provided insight into student thinking and as the project progressed, teachers reported having more conversations with students during learning and thus could respond to their needs.

With support and time, teachers can develop a good understanding of the Teaching as Inquiry process. This research has shown that with guidance, teachers not only changed their practices but also sustained this change over a long period.

In our view, the participating teachers, through their commitment to student learning, helped their students improve their ability to communicate science ideas and to improve their literacy and science literacy.

Application to Other Education Settings

The raising of the achievement of Māori and Pasifika students is a challenge nationally and there would be many schools that are trying to do this. Although the limitation of this study is that it is a case study of one school, what we have learnt could be applied in other similar contexts. The findings may also find application in teacher education where students are learning about teaching as inquiry and learning to teach science students with similar learning challenges.

We would like to thank the Ministry of Education for giving us the opportunity to do this research through the teacher-led innovation fund. We would also like to thank the teachers who took on the challenge of inquiring into their own practice and the students without whom this research would not have been possible.

Final Comment

Science is not done, it is not communicated, through verbal language alone. It cannot be. The 'concepts' of science are not solely verbal concepts, though they have verbal components. They are semiotic hybrids, simultaneously and essentially verbal, mathematical, visual-graphical, and actional—operational. The actional, conversational, and written textual genres of science are historically and presently, fundamentally and irreducibly, multimedia genres. To do science, to talk science, to read science and write science it is necessary to juggle and combine in various canonical ways verbal discourse, mathematical expression, graphical and –visual representation, and motor operations in the world (Lemke, 1998, p. 87).

References

Akcay, H. (2017). Constructing concept maps to encourage meaningful learning in science classroom. *Education, 138*(1), 9–16.

Baviskar, S. N., Hartle, R. T., & Whitney, T. (2008). Essential criteria to characterize constructivist teaching: Derived from a review of literature and applied to five constructivist-teaching method articles. *International Journal of Science Education, 1,* 1–10. https://doi.org/10.1080/095006907 01731121

Bybee, R. W., & Roberts, D. A. (2014). Scientific literacy, science literacy, and science education. In: Lederma, N. G., & Abell, S.K. (Eds.) *Handbook of Research on Science Education,* Vol. II (pp. 559–572). New York: Routledge.

Cervetti, G. N., & Hiebert, E. H. (2015). Knowledge, literacy, and the Common Core. *Language Arts, 92*(4), 256–269.

Darling-Hammond, L., Wei, R. C., Andree, A., Richardson, N., & Orphanos, S. (2009). *Professional Learning in the Learning Profession.* Washington, DC: National Staff Development Council .

Driver, R., & Bell, B. (1986). Students' thinking and the learning of science: A constructivist view. *School Science Review, 67*(240), 443–456.

Driver, R., Asoko, H., Leach, J., Scott, P., & Mortimer, E. (1994). Constructing scientific knowledge in the classroom. *Educational Researcher, 23*(7), 5–12.

Frankel, K. K., Becker, B. L., Rowe, M. W., & Pearson, P. D. (2016). From "what is reading?" to what is literacy? *Journal of Education, 196*(3), 7–17.

Goldschmidt, P. (2010). *Evaluation of Seeds of Science/Roots of Reading: Effective Tools for Development of Literacy Through Science in Early Grades.* Los Angeles, CA: National Center for Research on Evaluation, Standards, and Student Testing (CRESST).

Horsley, J. & Moeed, A. (2018). "'Inspire Me'-High-ability Students' Perceptions of School Science." *Science Education International* 29.3 (2018).

Lambotte, F., & Meunier, D. (2013). From bricolage to thickness: Making the most of the messiness of research narratives. *Qualitative Research in Organizations and Management: An International Journal, 8*(1), 85–100. https://doi.org/10.1108/17465641311327531

Leeds-Hurwitz, W. (2009). Social Construction of Reality. In S. Littlejohn, & K. Foss (Eds.), *Encyclopedia of Communication Theory* (pp. 892–895). Thousand Oaks, CA: Sage Publications. https://doi.org/10.4135/9781412959384.n344

Lemke, J. (1998). Multiplying meaning. *Reading Science: Critical and Functional Perspectives on Discourses of Science,* 87–113.

Ministry of Education. (2007). *The New Zealand Curriculum.* Wellington: Learning Media.

Moeed, A., & Easterbrook, M. (2016). Promising teacher practices: Students views about their science learning. *European Journal of Science and Mathematics Education, 4*(1), 17–24.

Moeed, A., & Hall, C. (2011). Teaching, Learning and Assessment of Science Investigation in Year 11: Teachers' Response to NCEA. *New Zealand Science Review, 68*(3), 95–102.

Osborne, J. (2015). Practical work in science: Misunderstood and badly used? *School Science Review, 96,* 357.

Palmer, D. H. (2009). Student interest generated during an inquiry skills lesson. *Journal of Research in Science Teaching, 46*(2), 147–165. https://doi.org/10.1002/tea.20263

Pearson, P. D., Moje, E., & Greenleaf, C. (2010). Literacy and science: Each in the service of the other. *Science, 328*(5977), 459–463.

Rennie, L. J., Feher, E., Dierking, L. D., & Falk, J. H. (2003). Toward an agenda for advancing research on science learning in out-of-school settings. *Journal of Research in Science Teaching, 40,* 112–120. https://doi.org/10.1002/tea.10067

Solomon, J. (1987). Social Influences on the Construction of Pupils' Understanding of Science. *Studies in Science Education, 14*(1), 63–82. https://doi.org/10.1080/03057268708559939

Talbot-Smith, M., Abell, S. K., Appleton, K., & Hanuscin, D. L. (Eds.). (2013). *Handbook of Research on Science Education.* London: Routledge .

Appendix A

Science Sparkler

SCIENCE STARTER FOR LEVELS 4 AND 5 OF THE NEW CURRICULUM

ENERGY

- Active energies
- Stored energies
- Measuring
- Energy resources
- Heat transfer

© The Author(s), under exclusive license to Springer Nature Singapore Pte Ltd. 2021
A. Moeed and B. Cooney, *Language Literacy and Science*,
SpringerBriefs in Education, https://doi.org/10.1007/978-981-16-4001-8

ACTIVITY: ACTIVE ENERGIES

WORDS

Active energy: Energy that is being used to do something like kinetic and heat.
Electrical energy: Flow of electrons in a metal wire.
Electron: A very tiny negatively charged
          particle found in every atom.
Heat energy: Causes particles to move faster.
Kinetic energy: Anything moving has this energy.
Sound energy: Energy produced by vibrating
          objects or things like vocal cords.
Vibrating: Moving back and forward.
Radiant energy: Energy that travels at the speed
          of light. A lot comes from the sun.

1.   Complete the following table using the clues to help you.

No.	Clue	Word
1	Energy that comes from the sun.	__ __ d i __ n __
2	If it's moving it has this type of energy.	__ i __ e __ __ c
3	Causes particles to move faster.	__ e __ t
4	The flow of electrons in metals.	__ l __ __ t r __ __ a __
5	It's noisy energy.	__ __ u __ __
6	Energy on the go.	__ c __ i __ __

2.   The first column of the table contains a list of jumbled words (anagrams). They are to do this activity and you are given their first letter. Unravel them then put them into the crossword grid. The first one has been done for you.

Jumbled word	Word
greenies	energies
itcave	a
tickine	k
unods	s
tracellice	e
uns	s
hate	h
pesty	t
nosei	n
celtoner	e
antraid	r

e	n	e	r	g	i	e	s

ACTIVITY: ENERGY CHECKPOINT

Use the information from the first 3 activities to complete these puzzles.

1.   Use a line to match up list A with list B to write a complete word. Write the completed word in the table on the right.

LIST A		LIST B
g r a v i t	•	• c a l
m a g	•	• i a n t
c h e m i	•	• i c
n u c	•	• n e t i c
r a d	•	• e t i c
h e	•	• a t i o n a l
e l a s t	•	• r i c a l
k i n	•	• l e a r
e l e c t	•	• u n d
s o	•	• a t

WORD

3.   Complete the following acrostic puzzle.

1.   Energy that causes particles to move faster.
2.   Electrical energy makes electrons move in __.
3.   Everything moving has ___ energy.
4.   Energy that is used to do something.
5.   Moving back and forward.
6.   Tiny negatively charged particle.
7.   Energy produced by vibrating objects.
8.   Energy that comes from the sun.
9.   Radiant energy travels at the ___ of light.

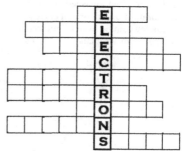

ACTIVITY: PUTTING IT ALL TOGETHER

Complete the crossword below. The words at the bottom of the page may be of use.

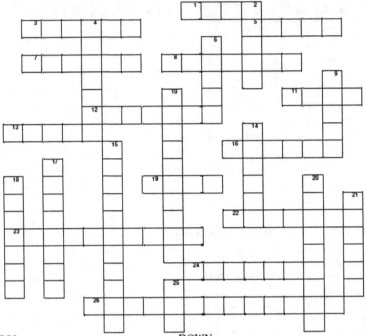

ACROSS

1. Its units are kilograms.
3. Mass describes this in an object.
5. Renewable energy source.
7. The ability to do work.
8. Energy produced from crops.
11. The basic building block of matter.
12. Energy contained in the atoms centre.
13. The unit of distance.
16. The measure of space liquid occupies.
19. This causes particles to move faster.
22. Energy when squashed or stretched.
23. Our most useful form of energy.
24. Energy coming from a light source.
26. Energy an aeroplane has in the air.

DOWN

2. Can turn turbines.
4. Generate electricity by spinning.
6. Comes from the sun.
9. The unit for energy.
10. Heated steam from the ground.
14. Fuels formed from buried plants.
15. The measure of how hot something is.
17. Energy important in electrical generation.
18. Anything moving has this energy.
20. Stored energy.
21. Is sound active or stored energy?
25. A fossil fuel.

Volume	Magnetic	Gravitational	Heat
Nuclear	Potential	Electrical	Atom
Radiant	Geothermal	Temperature	Solar
Metre	Steam	Mass	Oil
Active	Matter	Fossil	Tidal
Kinetic	Biomass	Joule	
Energy	Turbine	Elastic	

TOO EASY? TRY THESE

ACTIVITY: A CHILLY SITUATION

Why Don't Mountains get Cold?

Draw a line connecting each the dots by each word with its correct description. Circle any letters through which any line passes. Read down the circled letters to find the answer to the riddle. The first one has been done for you.

1   Travels as waves   • Chemical

2.   Made by vibrating objects   •   E   T
                                     H   H   • Elastic

3.   Possessed by all moving   •   • Gravitational
     objects   E

4.   Makes particles move   •   • Kinetic
     faster   Y

5.   Can't be created or   •   C • Magnetic
     destroyed   D   W

6.   Stored energy   •   I   E   • Heat

7.   Possessed by objects   •   A   • Radiant
     above the ground
                                  R

8.   Stored inside the nucleus   •   S   G   • Sound
     of an atom   S   L   D

9.   Types that are 'able to be   •   A
     seen'   N   • Electrical

10. Stored in bonds of   •   • Nuclear
     chemicals and food   O

11. Possessed by objects that   •   W   • Active
     are stretched   P   C

12. Moving charged particles   •   I   M   • Energy

13. Stored in certain metal   •   E   • Joule
     objects in a magnetic field A   A

14. Charged particles   •   F   • Potential

13. The unit of energy   •   P   • Solar

13. Renewable resource   • S   • Electrons

Answer_____

ACTIVITY: STRIKE A LIGHT

Why Was There Thunder and Lightning
in the Lab?

To find out, complete the acrostic puzzle. If there are 2 words in
the answer, don't leave a gap between them. When you have
finished, look at the letters in the column under the arrow.
Write them down in order from 1-25 and find out why. You may
need to use other sources for some answers.

1.   Unit for how hot or cold something is.

2.   Cells change chemical into __ energy.

3.   A sonic boom is this sort of energy.

4.   Energy alternative from the sea.

5.   Makes particles move faster.

6.   Magnetic is an example of ____ energy.

7.   Sound relies of this in the air.

8.   Renewable energy, expensive to harness.

9.   A renewable energy resource.

10. Produced in every energy conversion.

11. Living things need to ___ to get energy.

12. Energy inside batteries and bread.

13. Changes wind's energy into electrical.

14. Converts kinetic energy into to electrical.

15. Main energy supply in the sun.

16. Non-renewable energy resource.

17. A cake mixer changes electrical to __ energy.

18. An example of potential energy.

19. Gas produced in rubbish dumps.

20. Fossil fuel that produces pollution.

21. Future fuel to run cars.

22. An example of a renewable energy resource.

23. A TV remote changes electrical to __ energy.

24. Used to generate power in NZ.

25. Important part of radiant energy.

Answer: _____

Appendix B

Energy Word Quiz.

Please tick one answer in each of the following questions.

- **Which of these is a source of energy? ***
- Food
- Petrol
- Batteries
- All of the above

- **How many types of energy are there? ***
- 0
- 1
- 2
- More than 2

- **Which of these are types of energy? ***
- Sound
- Nuclear
- Elastic
- They all are

- **Conservation of energy means ***
- Energy can be created
- Energy can be destroyed
- Energy can be changed from one type to another
- All of the above

- **Thermal energy is another name for ***
- *Mark only one oval*
- Light energy
- Sound energy
- Heat energy
- Nuclear energy

- **A mobile phone transforms electrical energy into? ***
- Heat
- Light
- Sound
- All of the above

- **Which of these are types of energy? ***
- Sound
- Nuclear
- Elastic
- They all are

- **Conservation of energy means ***
- Energy can be created
- Energy can be destroyed
- Energy can be changed from one type to another
- All of the above

- **Thermal energy is another name for ***
- Light energy
- Sound energy
- Heat energy
- Nuclear energy

- **A mobile phone transforms electrical energy into? ***
- Heat
- Light
- Sound
- All of the above

- **Which of these is a source of energy? ***
- Food
- Petrol
- Batteries
- All of the above

- **How many types of energy are there? ***
- 0
- 1
- 2
- More than 2

(continued)

© The Author(s), under exclusive license to Springer Nature Singapore Pte Ltd. 2021
A. Moeed and B. Cooney, *Language Literacy and Science*,
SpringerBriefs in Education, https://doi.org/10.1007/978-981-16-4001-8

(continued)

● **Which of these is a source of energy?** *	● **Which of these are types of energy?** *
● Food	● Sound
● Petrol	● Nuclear
● Batteries	● Elastic
● All of the above	● They all are
● **Which of these can give us renewable energy** *	● **Conservation of energy means** *
● Wood	● Energy can be created
● Petrol	● Energy can be destroyed
● Wind	● Energy can be changed from one type to another
● Oil	● All of the above

Energy Keywords.

- **Which of these is a source of energy?** *

  - Food
  - Petrol
  - Batteries
  - All of the above

- **How many types of energy are there?** *

  - 0
  - 1
  - 2
  - More than 2

- **Which of these are types of energy?** *

  - Sound
  - Nuclear
  - Elastic
  - They all are

- **Conservation of energy means** *

- Energy can be created
- Energy can be destroyed
- Energy can be changed from one type to another
- All of the above

- **Thermal energy is another name for** *

  - *Mark only one oval.*
  - Light energy
  - Sound energy
  - Heat energy
  - Nuclear energy

- **A mobile phone transforms electrical energy into? ***

  - Heat
  - Light
  - Sound
  - All of the above

- **An energy transformation happens in a ***

  - Light bulb
  - Mobile phone
  - Toaster
  - All of the above

- **Food contains energy. The type of energy in food is called.. ***

  - Nuclear energy
  - Electrical energy
  - Sound energy
  - Chemical energy

- **Energy is measured in.. ***

  - Newtons
  - Joules
  - Kilograms
  - Watts

- **Potential energy is ***

  - energy that is stored up
  - energy that is doing something
  - both of the above

- **Which of these things has a lot of kinetic energy? ***

  - A car that is stopped
  - A bird that is flying
  - A person that is standing still
  - They all do
  - Send me a copy of my responses.

Powered by.

Screen reader support enabled.

Appendix C

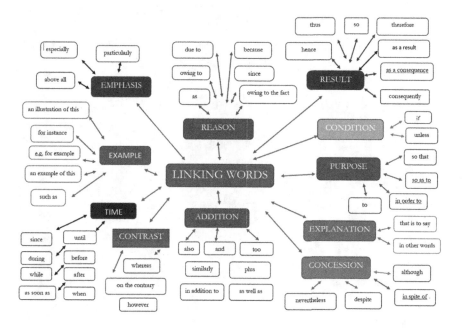

Printed in the United States
by Baker & Taylor Publisher Services

Printed in the United States
by Baker & Taylor Publisher Services